MOLECULAR SELF-ASSEMBLY IN NANOSCIENCE AND NANOTECHNOLOGY

Edited by **Ayben Kilislioğlu**
and **Selcan Karakuş**

Molecular Self-assembly in Nanoscience and Nanotechnology
http://dx.doi.org/10.5772/65607
Edited by Ayben Kilislioğlu and Selcan Karakuş

Contributors

Meizhen Yin, Baozhong Lü, Jinglin Shen, Xia Xin, Shiling Yuan, Lizong Dai, Conghui Yuan, Yiting Xu, Birong Zeng, Wei Zhang, Yin Zhao, Lu Yin, Galder Kortaberria, Maddalena Pedio, Barbara Ressel

Published by InTech
Janeza Trdine 9, 51000 Rijeka, Croatia

Publishing Process Manager Marijana Francetic
Technical Editor SPi Global
Cover InTech Design team

First published May, 2017
Printed in Croatia
Legal deposit, Croatia: National and University Library in Zagreb

Additional hard copies can be obtained from orders@intechopen.com

Molecular Self-assembly in Nanoscience and Nanotechnology, Edited by Ayben Kilislioğlu and Selcan Karakuş
p. cm.
Print ISBN 978-953-51-3157-1
Online ISBN 978-953-51-3158-8

Meet the editors

Professor Ayben Kilislioğlu is currently working in the Department of Chemistry, Istanbul University (IU), Turkey. She received her master of science degree in physical chemistry from IU in 1994. She received her doctor of philosophy degree in physical chemistry from IU in 2000. She worked as visiting research assistant professor at the University of Illinois, Chicago, Department of Chemistry, in the period 2005–2006. She also worked at the University of Chicago in Dr. Graeme Bell's Lab in 2007. She has research experience in molecular adsorption, surface characterization, and ion exchange. She has worked on different projects funded by the Istanbul University Grant Commission. She has published several research articles and a book chapter in this area.

Assistant professor Selcan Karakus is currently working in the Department of Chemistry, Istanbul University (IU), Turkey. She received her Master of Science degree in Physical Chemistry from IU in 2006. She received her Doctor of Philosophy degree in Physical Chemistry from IU in 2011. She worked as a visiting researcher at the University of Massachusetts, Department of Polymer Science and Engineering. She has research experience in molecular adsorption, self-assembled polymeric nanostructures and nanocomposites, and copolymer blends. She worked on different projects funded by Istanbul University. She has published several research articles and a book chapter in this area.

Contents

Preface

This book outlines studies about the self-assembly of several molecular species in modern chemistry that have improved our understanding of organized nanoscale structures. Self-assembly is a common principle in molecular fabrication of natural and synthetic systems and has important applications in nanoscience and nanotechnology. The formation of nanostructures by self-assembly occurs primarily through noncovalent interactions. Various nanostructures can be prepared through the different properties of noncovalent interactions. Well-defined two- and three-dimensional assemblies of molecular nanostructures have tremendous number of applications in areas of nanoscience and nanotechnology. In this book, readers will be able to elucidate beneficial examples and research results related to this field.

Ayben Kilislioğlu and Selcan Karakuş
Istanbul University, Istanbul, Turkey

Chapter 1

Nanostructured Morphologies by Self-Assembly of Diblock Copolymers: A Review

Galder Kortaberria

Additional information is available at the end of the chapter

http://dx.doi.org/10.5772/intechopen.68476

Abstract

Due to the thermodynamic incompatibility between blocks, diblock copolymers can self-assemble in a wide variety of nanostructures, covalent linkage among blocks preventing the phase separation at macroscopic scale. Those nanostructures depend on copolymer composition (f), Flory-Huggins interaction parameter among both blocks (χ), and polymerization degree of the copolymer (N). Thin films of block copolymers can show different equilibrium morphologies such as spheres, cylinders, gyroids, and lamellas. Besides mentioned parameters, film preparation process (substrate, annealing process if any) and used solvent will determine self-assembled morphology. In the present review, the most important morphologies or microstructures obtained for different diblock copolymer films are presented, as well as the most important phase transitions among them. Different microstructures and the way in which they can be obtained become of great importance, as they could be used as templates for nanoparticle deposition, nanolithography, or nanopatterned materials with several potential applications in different fields such as nanoelectronics or nanomedicine.

Keywords: self-assembly, nanostructure, morphology, lamellar, cylinders, gyroids

1. Introduction

Block copolymers consist in macromolecules produced by joining two or more chemically distinct polymer blocks, that may be thermodynamically incompatible. Segregation of these blocks on the molecular scale (5–100 nm) can produce many different complex nanostructures. More than three decades of theoretical development have culminated in remarkably predictive statistical theories that can account for the domain shapes, dimensions, connectivity and ordered symmetry of many types of block copolymers. Nowadays, the possibility to join blocks in novel molecular architectures can produce a really huge number of structured materials endowed with tailored mechanical, optical, electrical, barrier, and other physical properties [1, 2].

Many different block configurations can be constructed with different synthetic chemistry techniques. Based on the number of chemically distinct blocks, and their linear or branched sequencing linear and branched diblock copolymers, linear ABC triblock copolymers and ABC star or heteroarm triblock copolymers can be distinguished. Those would be the most important ones, even if block copolymers with more than three blocks can be also found.

The simplest and most studied architecture is the linear AB diblock, consisting of a long sequence of A type monomers covalently bonded to a chain of B type monomers. $(AB)_n$ multiblocks are formed by coupling additional A and B blocks. Morphologies obtained by self-assembly of diblock copolymers in films will be shown and reviewed in the present chapter.

Due to the thermodynamic incompatibility between blocks and the covalent bond among them that avoids macrophase separation, block copolymers can result in many different nanostructures [1, 3, 4]. For this reason, many researchers have shown interest in block copolymers during the last decade [5–7]. Block copolymers can self-assemble into different nanostructures that depend on several parameters such as copolymer composition (f), Flory-Huggins interaction parameter among both blocks (χ), and polymerization degree of the copolymer (N). Spheres, cylinders, gyroids, and lamellas are the usual equilibrium nanostructures shown by diblock copolymers either in bulk or in thin films [8–11] as it can be seen in **Figure 1**.

N and f, related to the translational and conformational entropy of the copolymer, respectively, can be tuned by polymerization reaction, while χ reflects enthalpic interactions among blocks. As a result, χN represents the enthalpy related to the linkage of two different polymer chains. In thin films of block copolymers, the orientation and lateral ordering of the microdomains are influenced by factors like χN, interfacial interactions at the air/polymer and polymer/substrate interfaces, and the commensurability between the equilibrium period (L_o) and film thickness [12, 13]. A parallel orientation of the cylindrical and lamellar microdomains is observed when there is a large difference in the surface energies of the two blocks and/or a strong preferential interaction of one block with the substrate. However, sometimes it is desirable to be able to orient the microdomains normal to the film surface. As the molar mass of the copolymer mainly determines the size of micro or nanodomains, it can be quite easily controlled by the polymerization process.

In the present chapter, a deep review on the generation of nanostructures by self-assembly of diblock copolymers in thin films is presented. Besides classical lamellar, cylindrical or spherical morphologies obtained for many different diblock copolymers, more complex nanostructures such as perforated lamellae or bicontinuous gyroids will be shown as obtained by several authors for many different diblock copolymer types.

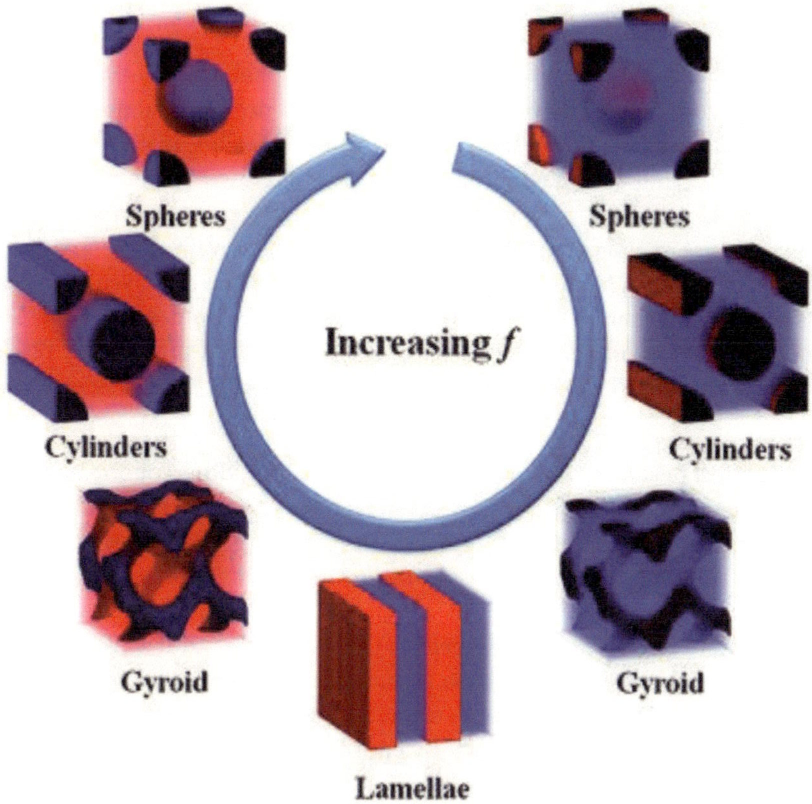

Figure 1. Various nanostructures formed by the self-assembly of block copolymers. The *f* indicates the volume fraction of one component. Reproduced with permission of Soft Matter 2013, 9, 9059. Copyright 2013, Royal Society of Chemistry.

2. Morphology of diblock copolymer thin films

Matsen and Schick [14] calculated the phase diagram for AB diblock copolymers. For $X_{AB}N < 10.5$, only a disordered melt is predicted. At larger values of $X_{AB}N$, above the order-disorder transition curve, five ordered microphase structures are predicted to have regions of thermodynamic stability. The lamellar (L) phase is stable for nearly symmetric diblocks, while a hexagonally packed cylinder (HPC) phase is stable for diblocks with intermediate levels of compositional asymmetry. With still more compositional asymmetry, the hexagonal phase gives way to a body-centered cubic spherical (BCC-sphere) phase. A very narrow region of close-packed spheres (CPS) separates the disordered and sphere phases at the composition extremes. Finally, Matsen and Schick predicted narrow regions of stability of a complex gyroid (G) phase close to the order-disorder transition and between the L and HPC phases. All those morphologies obtained for different block copolymers will be shown together with some others like perforated lamellae.

2.1. Lamellar morphology

It constitutes the stable phase for symmetric diblock copolymers. It has been commonly found for many diblock copolymer thin films, both without any further treatment or after thermal or solvent vapor annealing. Both thermal and solvent annealing methods have been used to bring the samples into their equilibrium states and to reduce the number of defects [15]. Lamellar morphology can be used to create line and space patterns on substrates by controlling their orientation. Lamellar orientation depends on several parameters and/or factors such as the substrate used for film preparation [16, 17], thermal or solvent vapor annealing treatments [15, 17, 18], chain length or molar mass [16, 17, 19], pressure application during annealing [20] or the use of graphoepitaxial techniques for film preparation [21–23]. Thus, different orientations have been obtained for different block copolymers by controlling those parameters.

Regarding the effect of substrate used for preparing the films, different orientations of lamellar microdomains have been found for poly(styrene-*b*-methyl methacrylate) (PS-*b*-PMMA) diblock copolymer by using random PS-r-PMMA copolymer-grafted substrates [16]. The control of styrene fraction in random copolymers allowed tuning the interfacial interactions at the substrates from PS-selective to PMMA-selective. Parallel orientations of lamellar microdomains were found for all substrates except in the case of a neutral substrate, for which no lamellar structure was obtained, attributed to surface compatibility between PS and PMMA blocks at the substrate. The importance of substrate on lamellar orientation and structure also has been seen for poly(ethylene oxide-*b*-buthylene oxide) (PEO-*b*-PBO) copolymer by preparing films both in mica and silicon substrates [17]. For films annealed under vacuum, a half-layered, densely branched structure was found for films prepared on mica. For films prepared on silicon, however, a half-layered structure without any meaningful feature was obtained. Probably, in the annealing treatment, the polymer chains located at the upper layers moved to the bottom ones. This makes the upper layers to progressively disappear. The thickness of the upper polymer layers became larger while that of the bottom layers in contact with mica decreased. An increase of lamellar thickness seemed to be the cause of both phenomena. In the first half layer, the increase of lamellar thickness led to spreading and half-layer thinning. In the upper layer, on the contrary, the islands contacted but got thicker. The interaction of PEO block with the substrate seemed to be reason for obtaining such a branched structure and differently oriented PEO crystals in the polymer layer in contact with mica. The effect of annealing process was also clearly seen, since as-cast films exhibited a multilayered structure with lamellar microdomains parallel to the surface for shorter copolymers, while for longer ones, the lamellar microdomains in the upper polymer layers exhibited mixed orientations of parallel and perpendicular lamellae.

Both thermal and solvent vapor annealing have also been used for the generation of lamellar morphologies with different orientations. In this way, the lamellar orientation of poly(styrene-*b*-butadiene) (PS-*b*-PB) copolymer has been found to be strongly dependent on solvent annealing process [15]. The reorientation of lamellae occurred during solvent vapor annealing with ethyl acetate. Films presented initially a distribution of lamellar orientations and the lamellar spacing depended on their orientation. Whereas the parallel lamellae presented the smallest spacing, the perpendicular ones had the largest one. Due to the effect of substrate that constrained perpendicular lamellae, their swelling was more difficult than that of the parallel ones. However, the measured lamellar spacing was equal for both after a certain time of swelling. The swelling process changed the orientation of the lamellae they appearing completely parallel at the end of the process. Schematic evolution of structures during solvent vapor annealing process can be seen in **Figure 2**.

Thermal annealing also affects the formation and orientation of lamellar domains. For poly (styrene-*b*-caprolactone) (PS-*b*-PCL) copolymer [18] annealed at temperatures higher than T_g of both blocks, worm-like or lamellar morphologies with lamellae parallel to the surface were obtained at 100 and 120°C, respectively, as it can be seen in **Figure 3**. Moreover, the addition of magnetic nanoparticles modified with polymeric brushes in order to increase compatibility with matrix did not change obtained morphologies or lamellar orientations.

To perform the annealing process under vacuum constitutes another way to control the formation and orientation of lamellar microdomains [17, 20]. Highly oriented films can be obtained by annealing copolymer films under pressure. In this way, for a lamella-forming poly(styrene-*b*-ethylene oxide) (PS-*b*-PEO) annealed under a pressure of 0.2 MPa, a highly oriented structure has been obtained by the combined effects of shear flow and self-organization of the block copolymer during annealing under stress [20]. Annealing at high temperature under pressure was carried out in two steps: the melt-flow of the copolymer in the X–Y plane under pressure and the ordering of the lamellar structure during annealing at high temperature. The degree of orientation increased with annealing time.

Chain length and molar mass of copolymers also are key factors in order to obtain lamellar structures with different orientations [16, 17, 19]. For films of PS-*b*-PMMA prepared in a neutral substrate [16], it was found that the orientation of lamellar microdomains was strongly influenced by the molecular weight of the copolymer and film thickness. For the copolymer with low molar molecular weight (29,000 g/mol), a parallel orientation of lamellar microdomains was obtained over the film. When the molecular weight was increased to 113,000 g/mol, a greater number of perpendicularly oriented lamellae were seen close to the polymer/substrate interface. By a theoretical approach, authors demonstrated that lamellar orientation on a neutral substrate is mainly determined by the entropic contribution to the free energy. For copolymers with higher N values, the chain stretching (nematic effect) entropically favored

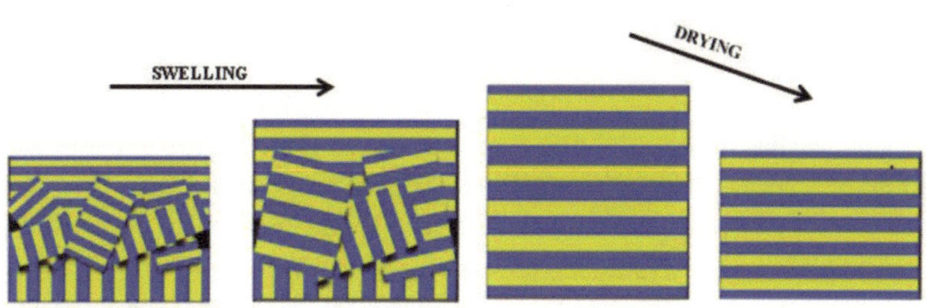

Figure 2. Schematic structures of the thin film during swelling and drying.

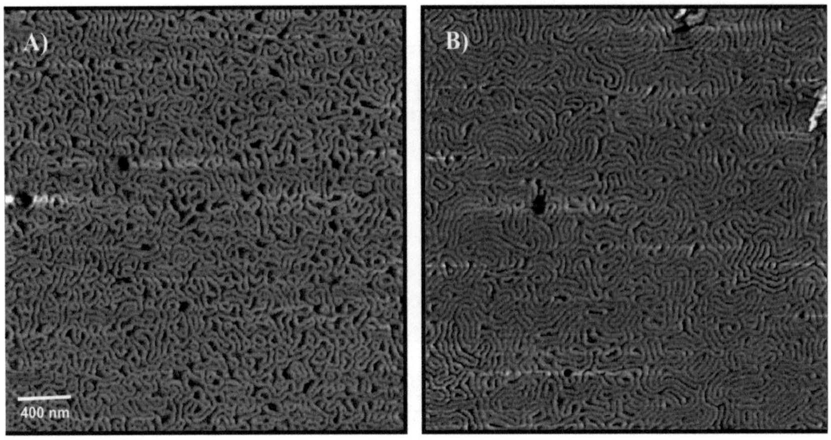

Figure 3. AFM phase images of PS-*b*-PCL block copolymer annealed for 72 h at: (A) 100°C and (B) 120°C. Reproduced with permission of [18]. Copyright 2014, Elsevier.

the perpendicular orientation of lamellae, while for copolymers with lower N values, the chain-end effect dominated and the parallel orientation of lamellae was favored. The effect of molecular weight on lamellar orientation was also found for PEO-*b*-PBO copolymer thin films prepared on mica [17]. For shorter copolymers, a multilayered lamellar structure parallel to the surface was found. In the half layer in contact with the surface, the stems of PEO crystals were parallel to it, while they appeared perpendicularly oriented in the upper half layer. In the other hand, the as-cast thin film of copolymers with higher molecular weights showed a mixture of parallely and perpendicularly oriented lamellae. The effect of chain length has also been studied for PS-*b*-PB copolymer thin films, together with that of film thickness [19], varying film thickness/lamellar thickness ratios from 0.5 to 10. For ratios between 1 and 10, lamellae paralelly oriented to the surface were obtained for low molar masses (below 55 kg/mol), while perpendicularly oriented lamellae were obtained for copolymers with higher molar masses (above 90 kg/mol). On the other hand, for film thicknesses equal to the lamellar thickness, the films did not exhibit any texture, whereas for film thicknesses equal to the half of lamellar thickness, a weak surface structure could be observed. Authors claimed that it was consistent with symmetric wetting (the same block adsorbs at both film interfaces), but with weak selectivity. For blocks that interact only weakly with the substrate surface and with air, molar mass was seen to be the key factor for the lamellar orientation. Authors gave an explanation of their findings based on the theoretical model by Pickett et al. [24]. High molar masses favored the perpendicular lamellar orientation, while lower ones resulted in less ordered parallel lamellae, due to the different scaling behavior of entropic and enthalpic contributions to the interfacial energies with chain length. By studying the effect of molar mass on lamellar structures, in poly(cyclohexylethylene-*b*-methyl methacrylate) (PCHE-*b*-PMMA) thin films, some of the smallest lamellae observed were found [25]. Ordered lamellar domain pitches were identified by small-angle X-ray scattering for copolymers containing 43–52 vol% of PCHE block. Atomic force microscopy was used to show around 7.5 nm lamellar features, some of the smallest observed. For the lower molar mass sample, sub-5 nm nanodomains were found, which together with the sacrificial properties of PMMA and the high overall thermal stability placed this material at the forefront of "high-χ" systems for advanced nanopatterning applications. Lamellae were found to be aligned perpendicularly to the surface.

Block copolymers have attracted considerable attention as fabrication method for nanopatterning, in which nanometer-sized domains are used as lithography templates. For dense line pattern formation, lateral lamellar domains are more preferable because in their case, an aspect ratio at least greater than 2 can be obtained. Epitaxial self-assembly by chemically or topographically nanopatterned substrates or resist patterns as guide have been demonstrated to be effective for lateral lamellar orientation [21–23]. In this way, lamellar domains of PS-*b*-PMMA symmetric copolymer have been successfully aligned on chemically nanopatterned substrates [23]. Advanced lithography allowed the induction of epitaxial self-assembly of domains in films. In this way, oriented patterns without defects were created over large areas. Obtained structures depend on the size and quality of the surface pattern rather than on the self-assembly process. **Figure 4** shows the two-step process used for the preparation of nanopatterned surfaces.

The lamellar microdomains of PS-*b*-PMMA followed the lithographically predefined surface patterns of chemical contrast. Bent block copolymer lamellae could be therefore obtained. Topographic guiding patterns have also been [21] used for bending lamellar microdomains, designed as elbows with varying corner angles. By controlling the surface of guiding patterns nonselective to the microdomains, the orientation of lamellae was rendered perpendicular to the surfaces of bottom and sidewalls of the guiding patterns. For both PS-*b*-PMMA and PS-*b*-PEO copolymers, authors demonstrated that bending of lamellar microdomains in block copolymer films could be formed within angled corners of topographic guiding patterns that exhibited a nonselective wetting property. Both lamellar-forming copolymers revealed similar bending trends. **Figure 5** shows the scheme of the procedure used to direct the self-assembly of block copolymers on topographic guiding patterns. Finally, also resist patterns have been used as guide for lamellar orientation [22]. In this way, the lateral alignment of symmetric PS-*b*-PMMA was achieved in confined spaces between straight guide patterns composed of a hydrogen silsesquioxane resist. It is worth to note that a lamellar structure with a period lower than that the pitch of the guide pattern was formed by this approach, while by chemically nanopatterned substrates, only structures with higher periods than the pitch of the guide pattern could be oriented.

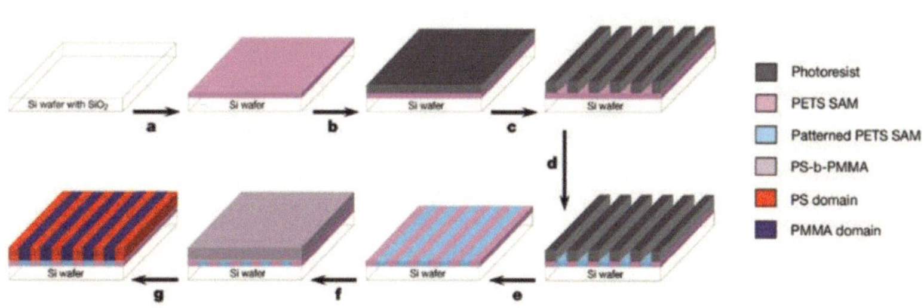

Figure 4. Two-step process for the preparation of nanopatterned surfaces for lamellar orientation of PS-*b*-PMMA copolymer. Reproduced with permission of [23]. Copyright 2003, Springer Nature.

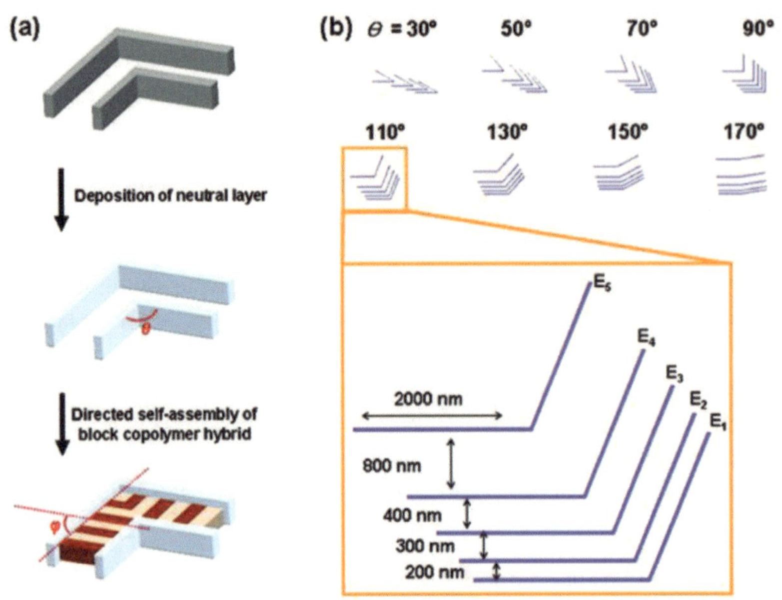

Figure 5. Schematic of the process used to direct the assembly of lamellae-forming block copolymers on topographic guiding patterns with bending geometries. (a) The procedure begins with the deposition of neutral layer followed by directed self-assembly of block copolymer thin films. (b) Diagram of topographic guiding patterns. Reproduced with permission of [31]. Copyright 2010, American Chemical Society.

2.2. Perforated lamellar morphology

Even if in the bulk state, the perforated lamellar morphology of block copolymers has been reported to be a metastable state of the gyroid (G) phase, under thin film confinement conditions, one-dimensional confinement provides stability to the phase [26, 27] and has been found for different diblock copolymer thin films under several conditions [26–35]. For thin films of poly(styrene-*b*-dimethylsiloxane) (PS-*b*-PDMS) diblock copolymer spin coated onto silicon wafers modified with hydroxyl-terminated PDMS subject to solvent vapor annealing, depending on the film thickness, its commensurability with the microdomain period, and the ratio of toluene/heptane vapors used for the solvent annealing process, perforated lamellae morphology can be obtained [26], besides spheres, cylinders, or gyroids. Even if PS-*b*-PDMS presented a double-gyroid morphology in bulk, as a thin film the morphology can be tuned by playing with different parameters. The solvent composition and pressure affected the relative Swelling of blocks, and therefore, the effective volume fraction, the swelled film thickness, and the effective χ. In-plane cylinders, HPL, and spheres with excellent long-range order and low defect levels were obtained at specific film thickness and solvent vapor compositions. For the same annealing conditions with 3:1 toluene/heptane ratio, hexagonally perforated lamellae (HPL) appeared and dominated as the film became thinner, as it can be seen in **Figure 6**.

At around 80 nm thickness, two layers of interconnected HPL were formed. Other solvent ratios produced mixed morphology structures such as comb-like patterns from coexisting

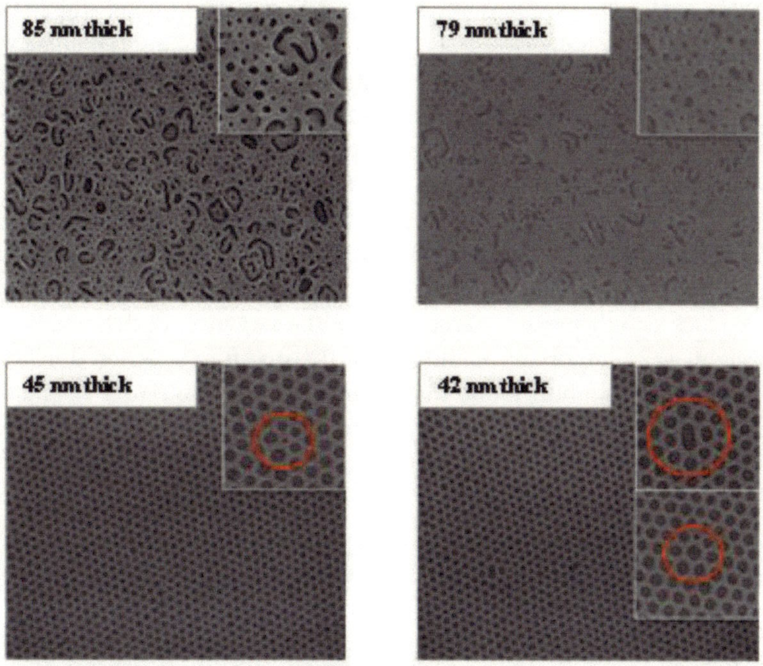

Figure 6. Representative SEM images of PDMS microdomains formed in thin films with differing initial thickness at the same solvent vapor annealing conditions. Reproduced with permission of [26]. Copyright 2014, American Chemical Society.

lamellae and cylinders. For thin film composites based on PS-*b*-PDMS copolymer with a volume fraction of 38% for PDMS and PS-modified Au nanoparticles (subjected to a thermally assisted solvent vapor treatment), an unique one-to-one positioning phenomenon of guest Au nanoparticles in the host microdomains was found; each of PS-functionalized nanoparticle appeared embedded within the perforation domain of HPL morphology of PS-*b*-PDMS. By controlling the weight fraction of Au-NPs and the effective volume fraction of constituent polymer blocks, individual nanoparticles were selectively incorporated into the centers of PS perforations of HPL morphology. The local minimization of free energy achieved by the placement of nanoparticles into the center of the perforation domain was theoretically supported by the self-consistent field theory (SCFT) simulation.

For poly(methyl methacrylate-*b*-vinyl-mtriphenylamine) (PMMA-*b*-PVmTPA) copolymers synthesized by RAFT polymerization with different PVmTPA fractions from 0.25 to 0.33, a variety of well-defined nanostructures were obtained [29]: HPC for a fraction of 0.27 and below, lamellae for fractions of 0.30 and above and HPL at an intermediate fraction of 0.29. The possibility of tuning morphologies obtaining different nanostructures for those block copolymers containing electro-active hole-transport components such as triphenylamine-based ones constitutes an attractive option for organic electronic applications in which well-defined nanoscale structures are desirable.

At a composition of 35 vol% of poly(methyl acrilate) (PMA), the formation of a HPL morphology was observed for a polydisperse poly(styrene-*b*-methyl acrilate) (PS-*b*-PMA) copolymer for short- and long-term solvent-casting conditions [30], as it can be seen in **Figure 7**. The hexagonal arrangement of the PS perforations is clearly visible in the upper right inset of the Figure which depicts a plane view of the PMA layer. The relative positions of the perforations furthermore indicated the ABC stacking of the PMA layers, characteristic for the rhombohedral structure. No order-order transitions were observed at elevated temperatures or after prolonged thermal annealing. The observed stabilization of the HPL morphology, considered to be metastable in narrow-disperse diblock copolymers as well as diblock copolymers with selective block polydispersity, suggested that the skewness of the distribution of block molecular weights is an important parameter for the structure selection during microphase separation. In particular, authors found that symmetric block molecular weight distributions made possible the stabilization of microstructures presenting higher standard deviation of mean curvature. Authors showed the importance of controlling the width and symmetry of

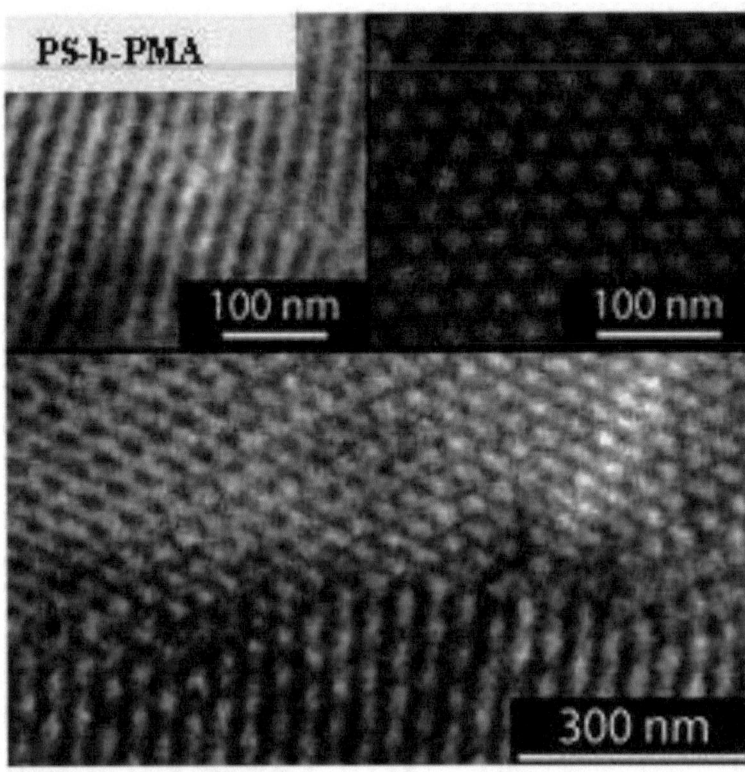

Figure 7. Bright field electron micrographs of PS-PMA samples after 72 h of thermal annealing at $T = 120°C$ and staining with RuO_4 (PMA is dark domain). $(PS\text{-}PMA)_{NP}$: revealing HPL microstructure imaged at low magnification. The inset on the left depicts the PS-perforations within the PMA layers. The inset on the right shows a plane view revealing the hexagonal arrangement of the PS perforations. Reproduced with permission of [30]. Copyright 2008, American Chemical Society.

molecular weight distribution for the design of different microstructures with special topological properties for different potential applications.

For poly(styrene-*b*-isoprene) (PS-*b*-PI) copolymer thin films spin coated on mica [31], the formation of HPL morphology was found for a volume fraction of 38% for PS. The HPL layers were oriented parallel to the substrate and the perforations also presented the ABC-type stacking structure. They claimed that the transition from HPL to double gyroid (DG) always started from the substrate side, being the development of DG {121} plane parallel to the substrate. This confirmed the epitaxial conversion among HPL layers and DG {121} plane. For PS-*b*-PI copolymer with 70% of PS, some of the thermal annealed samples presented HPL morphology. Epitaxial relations of phase transitions between the lamellar, HPL, and G morphologies were investigated by SAXS and rheology [32]. In HPL to G transitions, six spot patterns of G phase were observed in two-dimensional SAXS pattern. On the other hand, in direct L to G transition without appearance of HPL phase, the polydomain patterns of G phase were observed. Direct L to G transition of blend may be suppressed by high-energy barrier of transition and mismatches in domain orientation between epitaxially related lattice planes.

For the same copolymer but with 65% of PS [33], it was concluded that order-order transitions between the L, HPL, and G phases proceeded through nucleation and growth. Maintaining the continuity of microphase-separated interfaces across the resulting grain boundaries involved considerable local distortion of both morphologies. The different geometrical characteristics of the phases and the imbalance in the spacings of epitaxially related lattice planes constituted the reason for that. There was an increase of the surface energy of the grains, which strongly restricted nucleation, avoiding direct L to gyroid transitions. The formation of the metastable HPL structure under such conditions reflected the ease with which the L to PL transition could occur compared to L to G one. Similar effects dominate the G to L transition. Very similar conclusions were obtained by the same authors [33] for poly((ethylene-co-propylene)-*b*-dimethylsiloxane) (PEP-*b*-PDMS) and poly (ethylene oxide-*b*-ethylethylene) (PEO-*b*-PPE) copolymers with 64 and 72% of PEP and PEO, respectively.

For PS-*b*-PPE copolymer [34], a thermo reversible mesophase transition between low-temperature HPL and high-temperature HPC was found. The transformation process was accompanied by a 6% change in the principal spacing, yet oriented specimens retained their macroscopic orientation through the transition. The cylinder to HPL transformation proceeded on a time scale of tens of minutes. At deep undercoolings, the transformation rate slowed due to a reduction in molecular mobility. By contrast, the HPL to C transformation occurred nearly 2 orders of magnitude more rapidly, a difference that reflected the nature of the dominant fluctuation modes for the two structures.

Finally, for a poly(ethylene oxide-*b*-butylene oxide) (PEO-*b*-PBO) copolymer with 71% vol PEO [35], it was also found that the transition from the L to the G phase proceeded via an intermediate PL structure. It was observed that the G phase developed from an oriented HPL phase precursor as an isotropic distribution of grains. This indicated that growth of this structure did not occur epitaxially on a macroscopic scale.

2.3. Hexagonally packed cylinders

Considered as one of the stable morphologies for asymmetric diblock copolymers, it has been found for many different copolymers [36–52] as a result of their composition or by different annealing processes. In that way, for amphiphilic poly(4-di(9,9-dihexylfluoren-2-yl) styrene)-*b*-poly(2-vinylpyridine) (PStFl2m-*b*-P2VPn) [36] copolymer thin films self-assembled into different nanostructures as a function of copolymer block ratio and film preparation process: random two phases, horizontal hexagonal P2VP cylinders, and hexagonally close packed (HCP) P2VP spheres. Authors found that the hexagonal cylinder and HCP morphologies were not similar to those found from block copolymers with similar compositions. Well-developed nanostructures were obtained for films annealed under CS_2 vapors and subsequent thermal annealing. Surprisingly, the solvent-annealed 50/50 copolymer films, either with or without subsequent thermal annealing, formed in-plane–oriented hexagonal P2VP cylindrical structure rather than L structure that was expected from an equivalent block composition. 75/25 copolymer films annealed under CS_2 vapors, independently of posterior thermal annealing, self-assembled into HCP P2VP structure. As a cylindrical structure was expected from copolymer composition, authors pointed out that it could be due to an annulation of the volume fraction rule. Those morphologies were found to make influences on the electrical memory performances of the polymers. In particular, the switching-ON voltage was influenced by the nanostructures and the film layer thickness as well as by the composition.

HPC morphology has also been observed for thin films of poly(styrene-*b*-tert-butyl acrylate) (PS-*b*-PtBA) with 60 wt% of PS, in contrast with the L morphology observed for the copolymer with 35 wt% of PS [37]. HPC morphology has also been observed for thin films of methyl methacrylate and polyhedral oligomeric silsesquioxane (POSS)-functionalized methacrylate (PMMA-*b*-PMAPOSS) diblock copolymer after solvent vapor annealing with CS_2 and Post-thermal annealing [42]. The exposure to solvent vapors generated perpendicularly oriented PMMA cylinders hexagonally packed in the PMAPOSS matrix. Even if before thermal annealing both blocks remained amorphous, the annealing process led to the crystallization of POSS moieties.

Nonequilibrium morphologies were also found for poly(styrene-*b*-4vinylpiridine) (PS-*b*-P4VP) (50/50) [43] by using a supramolecule. The copolymer assembled within the framework generated by the supramolecule, obtaining nonequilibrium morphologies that the sole copolymer could not reach. Perpendicularly oriented HPC domains formed first by the self-assembly of the supramolecules, based on symmetric (PS-*b*-P4VP) and 3-pentadecylphenol (PDP) linked to 4VP by hydrogen bonds, as it is shown in **Figure 8**. After selective removal of ~90% of the PDP and a brief solvent annealing in a chloroform atmosphere, symmetric PS-*b*-P4VP, containing a trace amount of PDP, self-assembled forming hexagonal microdomains oriented normal to the surface.

For PS-*b*-PI (32/68) copolymer thin films annealed at 182°C for 1 day and subsequently quenched to room temperature to freeze the structures, patterns of double gyroid (DG), HPC, and their coexisting phases were found [44]. A new epitaxial transition path among both

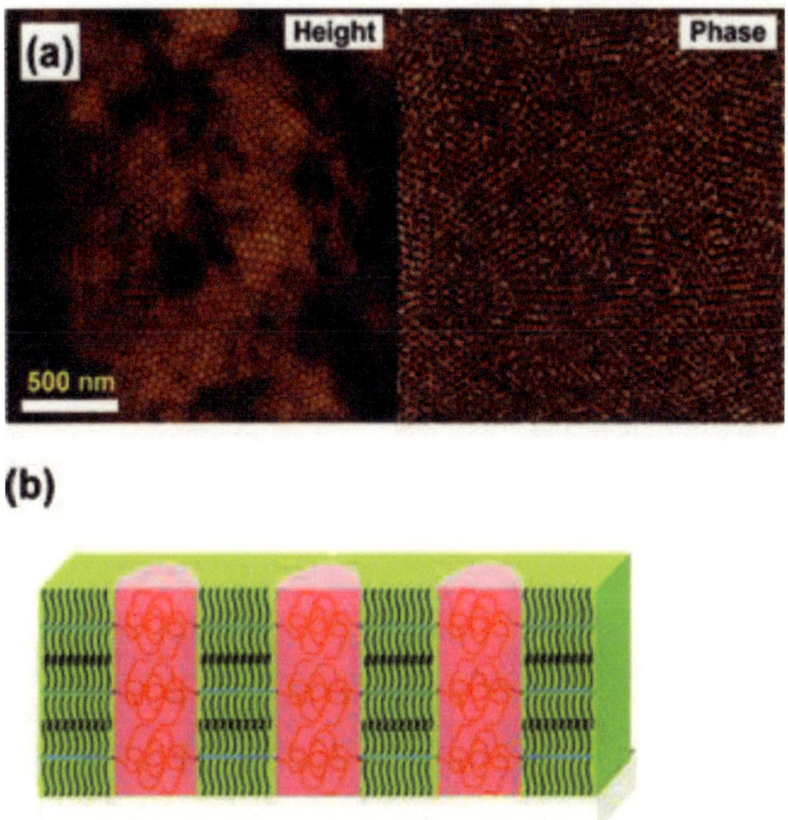

Figure 8. (a) SPM height and phase image of a PS(20,000)-*b*-P4VP(19,000)$(PDP)_1$ thin film, ~89 nm in thickness. The *z* scale is 15 nm for height and 30°for phase. (b) Schematic drawing of the cylinder-within-lamella hierarchical structure in thin films. Reproduced with permission of [43]. Copyright 2009, American Chemical Society.

nanostructures was found. This path seemed to have an advantage over the reported path in such a way that all the packed cylinders were converted from equivalent structures in DG while the previously reported path involved two different structures. In order to prepare nanoporous materials with periodic cylindrical holes, poly(butadiene-*b*-methylmethacrylate) (PB-*b*-PMMA) thin films that self-assembled into hexagonally packed PMMA cylinders in PB matrix were irradiated with γ-rays [45]. PMMA domains were removed by irradiation and succeeding solvent washing to form cylindrical holes within PB matrix, which was rigidified by radiation cross-linking: polymer nanoporous materials with periodic cylindrical holes were fabricated. For poly(ethylene oxide-*b*-styrene) (PEO-*b*-PS) copolymers with different block ratios, HPC [48] or inverse hexagonal cylinder (IHC) [47] morphologies have been found. For the 73/23 copolymer, IHC phase morphology was identified with PS HPC within the PEO matrix. Moreover, authors found that PEO blocks were tethered on the convex interfaces of the PS domains crystallizing outside of the cylinders. The orientation of PEO crystals changed with respect to the long cylinder axis due to the crystallization among PS cylinders. Authors pointed out that it only depended on crystallization temperature. For 37/63 copolymer, on the other side, parallely-oriented HPC morphology was obtained. Order-order transition was observed by dynamic mechanical analysis to occur through a C to G transition.

For a poly(styrene-*b*-(ethylene-alt-propylene)) (PS-*b*-PEP) copolymer [34] a thermoreversible mesophase transition was found between low-temperature HPL and high-temperature HPC, which occurred in various tens of minutes. Even if it increased rapidly with undercooling, for strong undercoolings, the rate decreased probably because of a decrease in molecular mobility. On the contrary, the HPL to HPC transformation occurred much more rapidly. For asymmetric poly(ethylene oxide-*b*-isoprene) (PEO-*b*-PI) copolymer [51], crystallization of PEO block from oriented HPC was investigated. It was found that crystallization of PEO from a shear-oriented HPC led to crystalline lamellar planes parallel to the cylinders and a step increase in domain spacing. HPC morphology has also been found for PMMA-*b*-PVmTPA copolymer with a volume fraction of 0.73 for PMMA [29] and for poly(isoprene-*b*-methylmethacrylate) (PI-*b*-PMMA) copolymer [52] thin films annealed under acetone vapors, as it can be seen in **Figure 9**. Moreover, in the last case, the morphology was maintained after the addition of modified magnetic nanoparticles. Obtained nanocomposites presented magnetic properties.

Figure 9. AFM phase images of PI-*b*-PMMA thin film annealed under acetone vapors solvent for 96 h. Reproduced with permission of [52]. Copyright 2016, Elsevier.

Liquid crystal (LC) block copolymers or copolymers containing a liquid crystal block has also been found to assemble into HPC structures [38, 39, 46, 49]. By working with a novel class of nanophase-separated and flexible double liquid crystalline EOBC-*b*-EOBCy diblock copolymers (EOBC and EOBCy monomers containing cyanobiphenyl mesogens) of different molecular weights but similar compositions [38], it was found that the diblock copolymer of higher molecular weight exhibited an exceptional order-order transition (OOT) from L to HPC upon nematic ordering. Obtained nanostructure type and size were determined by, besides the phase separation process, the flexibility of the polyether, and the mobility and order of the LC phases with different symmetries. Gopinadhan et al. [39] synthesized a novel liquid crystalline brush-like diblock copolymer, NBCB-*b*-NBPLA, using ring opening metathesis polymerization (ROMP), by sequential polymerization of functionalized norbornene (NB) monomers comprising cyanobiphenyl mesogens (CB) connected by 6 or 12 methylene units and poly(D,L-lactide) (PLA) chains of different lengths. They also used magnetic fields to direct the alignment of domains. By using copolymers with a NBPLA/NBCB ratio of 3:1, a morphology consisting on HPC of PLA was obtained. As the smectic mesophase presented diamagnetic anisotropy, the structure could be aligned by a magnetic field. HPC Domains of PLA minority block aligned paralelly to the direction of an imposed magnetic field were obtained. By UV cross-linking and selective etching of PLLA, it was possible the fabrication of thermally switchable polymer films with aligned nanopores. Alignment of self-assembled hierarchical microstructure in LC diblock copolymers using high magnetic fields was also achieved for ferroelectric liquid crystalline diblock copolymer of poly(styrene-*b*-isoprene-LC), (PS-*b*-PILC), incorporating a biphenyl 3-nitro-4-alkoxybenzoate LC mesogenic group and a non-LC block. HPC microstructure was successfully accomplished by application of a magnetic field at elevated temperatures. LC block containing cylinders lied parallel to the field, allowing for the production of thin films of vertical or "standing" cylinders. HPC morphology was also found for a LC-type amphiphilic poly(ethyleneoxide-*b*-methacrylic acid) with azobenzene moieties, PEOm-*b*-PMA(Az)n, where *m* and *n* stand for the degrees of polymerization of the polymer part and the LC part, respectively. Polymer films were submitted to thermal and barometric effects under hydrostatic pressure. Nanoscale structures exhibited an HPC of PEO cylinders arrangement. The role of interfacial interactions between hydrophilic PEO and hydrophobic PMA(Az) including mesogen sequences in the side chains, at the isotropic transition, was analyzed and found to be of crucial importance.

Electric field–induced alignment of microphase-separated block copolymer domains has also been carried out for different copolymers such as PS-*b*-PVP, PS-*b*-PI or PS-*b*-PB [50]. Upon applying an electric field to the film, a morphology consisting on hexagonally packed cylinders was obtained for all copolymers. The microdomains were oriented inside the cylinders. Authors pointed out that obtained orientation depended on two competing factors: interfacial interactions and electric field. The control of the interfacial interactions allowed the obtention of parallely or perpendicularly oriented cylinders.

HPC has also been found for several copolymers synthesized for potential biological applications [40, 41]. In that way, uracil-functionalized poly(3-caprolactone)-*b*-(4-vinylbenzyl uracil)s (PCL-*b*-PVBU) were synthesized for their bioinspired assembly forming different microstructures through nucleobase-induced supramolecular interactions [41]. The ordered morphologies of PCL-*b*-PVBU diblock copolymers changed from L to HPC or sphere morphologies with respect to the content of the hydrogen bond segment. Moreover, by hydrogen bonding between uracil and adenine groups, biocomplementary PCL-*b*-PVBU/9-hexadecyladenine (AC16), supramolecular complexes were formed. By using a nonstoichiometric PCL-*b*-PVBU/AC16 ratio, well-ordered L and HPC morphologies were obtained.

Finally, HPC has also been found for synthesized novel PEGylated polypeptide block copolymers of poly. The hierarchical self-assembly of their films led to the formation of L structures as a result of microphase separation of the diblock copolymers; HPC nanostructure featured α-helical conformations of PEGylated polypeptide segments, which were oriented perpendicularly to the director of the L structure formed by the diblock copolymers. These kinds of rod-coil block copolymers comprise an important class of nanomaterials because of their potential uses in biological and optoelectronic applications.

2.4. Gyroid morphology

The L and HPC morphologies cover a large range of compositions and molecular weights. In contrast, the gyroid (G) morphology corresponds to a narrow composition window at the weak segregation regime [2]. G morphology and transitions from lamellar or HPC morphologies to G ones have been found for different copolymers [32, 35, 44, 53–57], showing to have several potential applications.

For the production of devices as solar cells or special membranes, it results very interesting to obtain nanomaterials with connected pores. Thus, a way to prepare nanostructured porous materials would suppose a great advance for the design of such devices. As it seems to be difficult by common lithograpy, block copolymer lithography can be used as a "bottom up" approach, especially using DG structure. This morphology consists of two connected continuous networks of both blocks.

In that way, by using diblock copolymer-based PS-*b*-P4VP-(PDP) supramolecules that self-assemble into a bicontinuous gyroid morphology, metal nanofoams have been prepared [54]. G morphology consisted of PS network channels in a P4VP(PDP) matrix. After dissolving the PDP, the P4VP collapsed onto the PS struts and a bicontinuous gyroid interconnected template was obtained. The hydrophilic P4VP corona facilitated the penetration of water-based plating reagents into the porous template enabling a successful electroless metal deposition. Well-ordered inverse gyroid nickel foam was obtained after degrading the polymer by heating. Size and distribution of pores can be tuned by choosing the proper copolymer and PDP amount. As a candidate for the preparation of bicontinuous nanostructures for device fabrication, a poly(styrene-*b*-ferrocenylethylmethylsilane) (PS-*b*-PFEMS) diblock copolymer has been proposed [55]. The copolymer self-assembled into DG morphology. A block copolymer with a metal-containing block with iron and silicon in the main chain was selected due to its plasma etch resistance compared to the organic block, for further preparing nanoporous templates. Authors probed that DG morphology was the stable, equilibrium morphology of this copolymer.

The stability of G morphology at large segregation values was examined for poly(isoprene-*b*-ethylethylene) (PI-*b*-PEE) obtained by hydrogenation of PB block in PI-*b*-PB [57]. The large segregation values were achieved by the controlled and selective chemical modification of a precursor block copolymer, allowing χ to be varied over a large range at constant N. The complex bicontinuous gyroid structure was found to exist in a narrow window between the classical L and C morphologies. It was pointed out that the gyroid morphology remained even at segregation values (χN) much higher than those expected by theory and obtained experimentally. Authors performed annealing processes to the G phase, finding the same morphology and domain spacing. They obtained L and C structures by casting G samples with selective solvents. Samples were then thermally annealed, showing that G phase was again formed from cast phase. This seemed to indicate that G phase appeared as stable phase in the strong segregation regime.

A successful application of an ordered bicontinuous gyroid semiconducting network obtained from block copolymer template in a hybrid bulk heterojunction solar cell has been reported [56]. The freestanding G network was fabricated by electrochemical deposition into the 10 nm wide voided channels of a self-assembled, poly(4-fluorostyrene-*b*-lactide) (PFS-*b*-PLA) selectively degradable block copolymer film. Authors showed that the highly ordered pore structure was ideal for uniform infiltration of an organic hole transporting material, and solid-state dye-sensitized solar cells exhibited up to 1.7% power conversion efficiency. This patterning technique can be readily extended to other promising heterojunction systems consisting of a major step toward realizing the full potential of self-assembly in the next generation of device technologies. Schematic representation of the whole process can be seen in **Figure 10**.

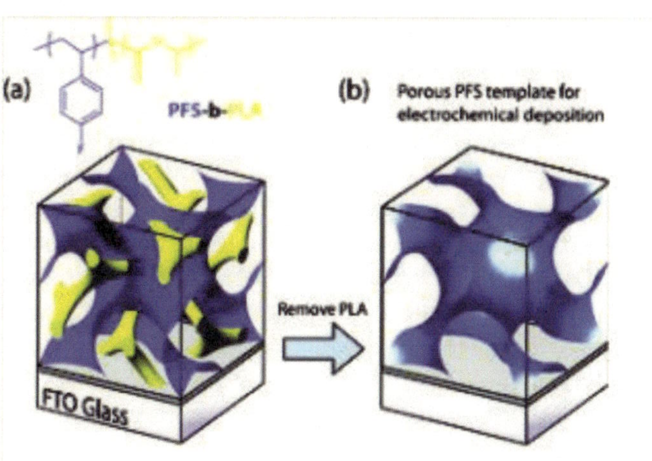

Figure 10. Schematic representation of the generation of porous PFS template for electrochemical deposition: a) preparation of PFS-*b*-PLA film onto FTO glass substrate and b) removal of PLA block. Reproduced with permission of [56]. Copyright 2009, American Chemical Society.

The epitaxial relations of phase transitions between the L, HPL and G morphologies were investigated for PS-*b*-PI (70/30) copolymer [32]. HPL to G transition was observed together with direct L to G one without appearance of HPL. Other authors also found the same transition for the same copolymer but with 38% of PS. In this case, films were prepared in mica substrate [31]. For thin films of PS-*b*-PI copolymer with 68% PI, however, a new epitaxial phase transition was found between DG and HPC structures [44]. The direction of phase transition, DG to HPC, was the main novelty of the work. A new epitaxial phase transition path from DG to HPC was described, which involved epitaxial relations between $\{121\}_{DG}$ and $\{10\}_{HPC}$ and between $\{111\}_{DG}$ and $\{11\}_{HPC}$. $\{220\}_{DG}$ became the grain boundary plane. The previously cited L to HPL to G transitions have also been found for films from other copolymers such as PEO-*b*-PBO [35], PEP-*b*-PDMS or PEO-*b*-PPE [33] prepared by melt-pressing. Their mechanisms have been well explained.

Finally, the ordered bicontinuous double diamond (OBDD) structure, that has long been believed to be an unstable ordered network nanostructure, relative to the ordered bicontinuous double gyroid (OBDG) structure for diblock copolymers, has been found for a copolymer composed of a stereoregular block, syndiotactic poly(propylene-*b*-styrene) (sPP-*b*-PS) [53]. The OBDD structure underwent a thermally reversible order-order transition (OOT) to OBDG upon heating. The thermodynamic stability of the OBDD structure was attributed to the ability of the configurationally regular sPP block to form helical segments, even above its melting point.

2.5. Spheres

Spherical morphology can consist mainly on body-centered cubic spheres (BCC), hexagonally close packed spheres (HP) or face-centered cubic spheres (FCC). Compared with the L and C nanostructures, the spheres of the BCC structure do not have anisotropy, and thus, the orientation of the BCC structure in thin films is in general less complicated than that of the HPC or L structures [58]. Morphology consisting on spheres has been found for several copolymers [36, 58–63]. Regarding to BCC morphology, it has been found for thin films of asymmetric poly(dimethylsiloxane-*b*-{2,5-bis[(4-butoxyphenyl)oxycarbonylstyrene}) (PDMS-*b*-PBPCS) (volume fraction of PDMS ranging from 10 to 23%) copolymer thermally annealed [58]. For all films, domains with a rectangular unit cell similar to the projection of the BCC lattice along the [110] direction were obtained. The orientation, parallel to the substrate, seemed to maximize the presence of PDMS at the substrate and at free interfaces. The surface layers of thin films were composed of both PDMS and PBPCS blocks. For copolymers with PDMS volume fractions of 10% and 13%, the neighboring [110]-oriented BCC domains matched well with each other, being the boundaries defect-free. For the copolymer with 23% of PDMS, PDMS spheres in the [110]-oriented BCC domains appeared overlapped with each other.

The mechanism and process of the thermally induced OOT of PS-*b*-PI diblock copolymer thin films from HPC morphology to BCC spheres have also been investigated [59]. Even if some studies dealing with the same kind of OOT have been reported [64, 65], authors investigated experimentally the epitaxial process by using a Bonse-Hart-type USAX apparatus [59], elucidating the mechanism and process of the thermally induced OOT.

BCC-sphere morphology has also been found for highly asymmetric poly(styrene-*b*-methacrylic acid) (PS-*b*-PMAA) copolymer [60]. The copolymer microphase-separated into spherical PMAA microdomains in a BCC arrangement in a PS matrix, in the strong segregation regime. It was found that the spherical microdomain morphology (with BCC symmetry) was favored over HPC morphology. BCC-type morphology has also been found for LC copolymers [61] such as poly(ethylene oxide)-*b*-(11-[4-(4-butylphenylazo)phenoxy]undecyl methacrylate)) (PEO-*b*-PMAAZ) one. For thermally annealed thin films, temperature-dependent phase transition behavior and the resulting morphologies were investigated. It formed a BCC structure of spherical PEO domains in the PMAAZ matrix, with both blocks in the amorphous state. The BCC structure was converted to HPC near 120°C, induced by the transformation of the isotropic phase of the PMAAZ matrix to the smectic A phase, composed of a laterally ordered structure of PMAAZ blocks with fully extended side groups. The resulting HPC structure was found to be very stable below 120°C, being retained as the smectic A phase of the PMAAZ matrix underwent further transitions to smectic C and to a smectic X phases.

Apart from BCC, other sphere symmetries have also been found in diblock copolymers [61–63]. For thin films of amphiphilic poly(4-di(9,9-dihexylfluoren-2-yl)styrene)-*b*-(2-vinylpyridine)) ($PStFl_{2m}$-*b*-$P2VP_n$) copolymer with a weight fraction of 73% for PStFl2 block, annealed under CS_2 vapors, a morphology consisting on hexagonally close packed (HCP) P2VP spheres was obtained rather than cylindrical structure which was expected from an asymmetric block composition [36], while for other compositions and annealing procedures, horizontal hexagonal P2VP cylinders or random two phases were found. The existence of FCC-packed spheres has also been probed for PEO-*b*-PB copolymer in the melt state [63]. The generation of FCC lattice took place by stacking together (111) planes in ABCABC sequence. It was noted that the FCC phase developed only by cooling from the disordered micellar phase.

Acknowledgements

Financial support from the Basque Country Government (Grupos Consolidados, IT-776–13) and the Ministry of Economy and Competitiveness (MAT2015–66149-P)-Spain is gratefully acknowledged.

Author details

Galder Kortaberria

Address all correspondence to: galder.cortaberria@ehu.eus

"Materials + Technologies" Group, University of the Basque Country (UPV/EHU), Donostia, Spain

References

[1] Bates FS, Fredrickson GH. Block copolymers-designer soft materials. Physics Today. 1999;**52**:32–38. DOI: 10.1063/1.882522

[2] Hamley IW. The Physics of Block Copolymers. Oxford University Press, Oxford, UK. 1999. p.432

[3] Leibler L. Theory of microphase separation in block copolymers. Macromolecules. 1980;**13**:1602–1617. DOI: 10.1021/ma60078a047

[4] Hadjichristidis N, Pispas S, Fluodas GA. Block copolymers: Synthetic strategies, physical properties, and applications. Wiley-interscience, Hoboken, USA. 2003

[5] Luo M, Epps TH. Directed block copolymer thin film self-assembly: Emerging trends in nanopattern fabrication. Macromolecules. 2013;**46**:7567–7579. DOI: 10.1021/ma401112

[6] Singh AN, Thake RD, More JC, Sharma PK, Agrawal JC. Block copolymer nanostructures and their applications: A review. Polymer-Plastics Technology and Engineering. 2015;**54**:1077–1095. DOI: 10.1080/03602559.2014.986811

[7] Yoo HG, Byun M, Jeong CK, Lee KJ. Performance enhancement of electronic and energy devices via block copolymer self-assembly. Advanced Materials. 2015;**27**:3982–3998. DOI: 10.1002/adma.201501592

[8] Balsara NP. Kinetics of phase transitions in block copolymers. Current Opinion in Solid State & Material Science. 1999;**4**:553–558. DOI: 10.1016/S1359-0286(00)00012-7

[9] Castelletto V, Hamley IW. Morphologies of block copolymer melts. Current Opinion in Solid State & Material Science. 2004;8:426–438. DOI: 10.1016/j.cossms.2005.06.001 DOI:10.1016/j.cossms.2005.06.001#doilink

[10] Matsen MW, Bates FS. Unifying weak- and strong-segregation block copolymer theories. Macromolecules. 1996;**29**:1091–1098

[11] Fasolka MJ, Mayes AM. Block copolymer thin films: Physics and applications. Annual Review of Material Research. 2001;31:323–355. DOI: 10.1146/annurev.matsci.31.1.323

[12] Nie T, Zhao Y, Xie Z, Wu C. Micellar formation of poly(caprolactone-*block*-ethylene oxide-*block*-caprolactone) and its enzymatic biodegradation in aqueous dispersion. Macromolecules. 2003;**36**:8825–8829. DOI: 10.1021/ma035131

[13] He CL, Sun JR, Deng MX, Chen XS, Jing XB. Study of the synthesis, crystallization, and morphology of poly(ethylene glycol)–poly(ε-caprolactone) diblock copolymers. Biomacromolecules. 2004;**5**:2042–2047. DOI: 10.1021/bm049720e

[14] Matsen MW, Schick M. Stable and unstable phases of a diblock copolymer melt. Physical Review Leters. 1994;**72**:2660–2663. DOI: 10.1103/PhysRevLett.72.2660

[15] Zhang J, Posselt D, Smilgies DF, Perlich J, Kyriakos K, Jaksch S, Papadakis CM. Lamellar diblock copolymer thin films during solvent vapor annealing studied by GISAXS: Different behavior of parallel and perpendicular lamellae. Macromolecules. 2014;**47**: 5711–5718. DOI: 10.1021/ma500633b

[16] Choi S, Kim E, Ahn H, Naidu S, Lee Y, Ryu DY, Hawker CJ, Russell TP. Lamellar microdomain orientation and phase transition of polystyrene-*b*-poly(methyl methacrylate) by controlled interfacial interactions. Soft Matter. 2012;**8**:3463–3469. DOI: 10.1039/c2sm07297a

[17] Liang GD, Xu JT, Fao ZQ, Mai SM, Ryan AJ. Morphology of semicrystalline oxyethylene/oxybutylene block copolymer thin films on mica. Polymer. 2007;**48**:7201–7210. DOI: 10.1016/j.polymer.2007.09.049

[18] Barandiaran I, Cappelletti A, Strumia M, Eceiza A, Kortaberria G. Generation of nanocomposites based on (PMMA-PCL)-grafted Fe_2O_3 nanoparticles and PS-*b*-PCL block copolymer. European Polymer Journal. 2014;**58**:226–232. DOI: 10.1016/j.eurpolymj.2014.06.022

[19] Busch P, Posselt D, Smilgies DM, Rheinlalnder B, Kremer F, Papadakis CM. Lamellar diblock copolymer thin films investigated by tapping mode atomic force microscopy: Molar mass dependence of surface ordering. Macromolecules. 2003;**36**:8717–8727. DOI: 10.1021/ma034375r

[20] Li Y, Kaito A. Highly oriented structure formed in a lamella-forming diblock copolymer with high molar mass. European Polymer Journal. 2006;**42**:1986–1993. DOI: 10.1016/j.eurpolymj.2006.04.007

[21] Pickett GT, Witten TA, Nagel SR. Equilibrium surface orientation of lamellae. Macromolecules. 1993;**26**:3194–3199. DOI: 10.1021/ma00064a033

[22] Park SM, Dong M, Rettner CT, Dandy DS, Wang Q, Kim HC. Bending of lamellar microdomains of block copolymers on non-selective surfaces. Macromolecules 2010;**43**:1665–1670. DOI: 10.1021/ma9020196

[23] Yamaguchi T, Yamaguchi H. Resist-pattern guided self-assembly of symmetric diblock copolymer. Journal of Photopolymer Science and Technology. 2006;**19**:385–388. DOI: 10.2494/photopolymer.19.385

[24] Kim SO, Solak HH, Stoykovich MP, Ferrier NJ, de Pablo JJ, Nealey PF. Epitaxial self-assembly of block copolymers on lithographically defined nanopatterned substrates. Nature. 2003;**424**:411–414. DOI: 10.1038/nature01775

[25] Kennemur JG, Yao L, Bates FS, Hillmyer MA. Sub-5 nm domains in ordered poly (cyclohexylethylene-b-methyl methacrylate) block copolymers for lithography. Macromolecules. 2014;**47**:1411–1418. DOI: 10.1021/ma4020164

[26] Bai WB, Hannon AF, Gotrik KW, Choi HK, Aissou K, Liontos G, Ntetsikas K, Alexander-Katz A, Avgeropoulos A, Ross CA. Thin film morphologies of bulk-gyroid polysty-

rene-block-polydimethylsiloxane under solvent vapor annealing. Macromolecules. 2014;**47**:6000–6008. DOI: 10.1021/ma501293n

[27] Heckmann M, Drossel B. Strong stretching theory for diblock copolymers in thin films. Journal of Chemical Physics. 2008;**129**:214903. DOI: 10.1063/1.3027437

[28] Nam TW, Jeong JW, Choi MJ, Baek KM, Kim JM, Hur YH, Kim YJ, Sik Jung YS. Single nanoparticle localization in the perforated lamellar phase of self-assembled block copolymer driven by entropy minimization. Macromolecules. 2015;**48**:7938–7944. DOI: 10.1021/acs.macromol.5b01931

[29] Mastroianni SE, Patterson JP, O'Reilly RK, Epps TH. Poly(methyl methacrylate-block-vinyl-mtriphenylamine): Synthesis by RAFT polymerization and self-assembly. Soft Matter. 2013;**9**:10146–10154. DOI: 10.1039/c3sm51806j

[30] Listak J, Jakubowski W, Mueller L, Plichta A, Matyjaszeski K, Bockstaller, MR. Effect of symmetry of molecular weight distribution in block copolymers on formation of "metastable" morphologies. Macromolecules. 2008;**41**:5919–5927. DOI: 10.1021/ma800816j

[31] Park HW, Im K, Chung B, Ree M, Chang T, Sawa K, Jinnai H. Direct observation of HPL and DG structure in PS-*b*-PI thin film by transmission electron microscopy. Macromolecules. 2007;**40**:2603–2605. DOI: 10.1021/ma062826c

[32] Ahn JH, Zin WC. Mechanism of morphological transition from lamellar/perforated layer to gyroid phases. Macromolecular Research. 2003;**11**:152–156. DOI: 1598–5032/06/152–05

[33] Hajduk DA, Ho RM, Hillmyer MA, Bates FS, Almdal K. Transition mechanisms for complex ordered phases in block copolymer melts. Journal of Physical Chemistry B. 1998;**102**:1356–1363. DOI: S1089–5647(97)02871-X

[34] Lai C, Loo YL, Register RA. Dynamics of a thermoreversible transition between cylindrical and hexagonally perforated lamellar mesophases. Macromolecules. 2005;**38**:7098–7104. DOI: 10.1021/ma050953n

[35] Hamley IW, Fairclough JPA,Ryan AJ, Mai SM,Booth C. Lamellar-to-gyroid transition in a poly(oxyethylene).poly(oxybutylene) diblock copolymer. Physical Chemistry Chemical Physics. 1999;**1**:2097–2101. DOI: 10.1039/A807847E

[36] Wi D, Ree BJ, Ahn B, Hsu JC, Kim J, Chen WC, Ree M. Structural details and digital memory performances of difluorene-containing diblock copolymers in nanoscale thin films. European Polymer Journal. 2016;**81**:582–597. DOI: 10.1016/j.eurpolymj.2015.12.011

[37] Pierre Escale P, Maud Save M, Billon L, Ruokolainenb J, Rubatat L. When block copolymer self-assembly in hierarchically ordered honeycomb films depicts the breath figure process. Soft Matter. 2016;**12**:790–797. DOI: 10.1039/c5sm01774b

[38] Wei W, Liu Y, Xiong H. Hierarchical nanostructures and self-assemblies in smectic-nematic liquid crystalline diblock copolymers. ACS Macroletters. 2014;**3**:892–895. DOI: 10.1021/mz500460j

[39] Gopinadhan M, Deshmukh P, Choo Y, Majewski PW, Bakajin O, Elimelech M, M. Kasi RM, Osuji CO. Thermally switchable aligned nanopores by magnetic-field directed self-assembly of block copolymers. Advanced Materials. 2014;**26**:5148–5154. DOI: 10.1002/adma.201401569

[40] Li PC, Lin YC, Chen M, Kuo SW. Self-assembled structures from PEGylated polypeptide block copolymers synthesized using a combination of ATRP, ROP, and click chemistry. Soft Matter. 2013;**9**:11257–11269. DOI: 10.1039/c3sm52061g

[41] Lin H, Cheng CC, Chuang WT, Chen JK, Jeng US, Ko FH, Chu CW, Huang CF, Chang FC. Bioinspired assembly of functional block-copolymer nanotemplates. Soft Matter. 2013;**9**:9608–9614. DOI: 10.1039/c3sm51870a

[42] Jin S, Hirai T, Ahn B, Rho Y, Kim KW, Kakimoto M, Gopalan P, Hayakawa T, Ree M. Synchrotron grazing incidence X-ray scattering study of the morphological structures in thin films of a polymethacrylate diblock copolymer bearing POSS moieties. Journal of Physical Chemistry B. 2010;**114**:8033–8042. DOI: 10.1021/jp1008785

[43] Tung SH, Xu T. Templated assembly of block copolymer toward non-equilibrium nanostructures in thin films. Macromolecules. 2009;**42**:5761–5765. DOI: 10.1021/ma900497j

[44] Park HW, Jung J, Chang T, Matsunaga K, Jinnai H. New epitaxial phase transition between DG and HEX in PS-b-PI. Journal of the American Chemical Society. 2009;**131**:46–47. DOI: 10.1021/ja808259m

[45] Sekine R, Sato N, Matsuyama T, Akasaka S, Hasegawa H. Radiation-induced fabrication of polymer nanoporous materials from microphase-separated structure of diblock copolymers as a template. Journal of Polymer Science: Part A: Polymer Chemistry. 2007;**45**:5916–5922

[46] Boyer SA, Grolier JPE, Yoshida H, Iyoda T. Effect of interface on thermodynamic behavior of liquid crystalline type amphiphilic diblock copolymers. Journal of Polymer Science: Part B: Polymer Physics. 2007;**45**:1354–1364. DOI: 10.1002/polb

[47] Huang P, Zheng JX, Leng S, Van Horn RM, Jeong KU, Guo Y, Quirk RP, Cheng SZD, Lotz B, Thomas EL, Hsiao BS. Poly(ethylene oxide) crystal orientation changes in an inverse hexagonal cylindrical phase morphology constructed by a poly(ethylene oxide)-block-polystyrene diblock copolymer. Macromolecules. 2007;**40**:526–534. DOI: 10.1021/ma061871h

[48] Mao H, Hillmyer MA. Macroscopic samples of polystyrene with ordered three-dimensional nanochannels. Soft Matter. 2006;**2**:57–59. DOI: 10.1039/b513958a

[49] Osuji C, Ferreira PJ, Mao G, Ober CK, Vander Sande JB, Thomas EL. Alignment of self-assembled hierarchical microstructure in liquid crystalline diblock copolymers using high magnetic fields. Macromolecules. 2004;**37**:9903–9908. DOI: 10.1021/ma0483064

[50] Xiang H, Lin Y, Russell TP. Electrically induced patterning in block copolymer films. Macromolecules 2004;**37**:5358–5363. DOI: 10.1021/ma049888s

[51] Hamley IW, Castelletto V, Floudas G, Schipper F. Templated crystallization from oriented gyroid and hexagonal melt phases in a diblock copolymer. Macromolecules. 2002;**35**:8839–8845. DOI: 10.1021/ma0207069

[52] Barandiaran I, Grana E, Katsiggianopoulos D, Avgeropoulos A, Kortaberria G. Nanocomposites based on nanostructured PI-b-PMMA and selectively placed PMMA-modified magnetic nanoparticles: Morphological and magnetic characterization. European Polymer Journal. 2016;**75**: 514–524. DOI:10.1016/j.eurpolymj.2016.01.005

[53] Chu CY, Jiang X, Jinnai H, Pei RY, Lin WF, Tsai JC, Chen HL. Real-space evidence of the equilibrium ordered bicontinuous double diamond structure of a diblock copolymer. Soft Matter. 2015;**11**:1871–1876. DOI: 10.1039/c4sm02608j

[54] Vukovic I, Punzhin S, Vukovic Z, Onck P, De Hosson JTM, Brinke GT, Loos K. Supramolecular route to well-ordered metal nanofoams. ACS Nano. 2011;**8**:6339–6348. DOI: 10.1021/nn201421y

[55] Gwyther J, Lotze G, Hamley I, Manners I. Double-gyroid morphology of a polystyrene-block-poly(ferrocenylethylmethylsilane) diblock copolymer: A route to ordered bicontinuous nanoscale architectures. Macromolecular Chemistry and Physics. 2011; **212**:198–202. DOI: 10.1002/macp.201000496

[56] Crossland EJW, Kamperman M Nedelcu M, Caterina Ducati C, Wiesner U, Smilgies DM, GES, Hillmyer MA, Ludwigs S, Steiner OU, Snaith HJ. A bicontinuous double gyroid hybrid solar cell. Nano Letters. 2009;**9**:2807–2812. DOI: 10.1021/nl803174p

[57] Davidock DA, Hillmyer MA, Lodge TP. Persistence of gyroid morphology at strong segregation in diblock copolymers. Macromolecules. 2003;**36**:4682–4685. DOI: 10.1021/ma034364y

[58] Shi LY, Li H, Lei WW, Ni W, Ran R, Pan Y, Fan XH, Shen Z. Extraordinary boundary morphologies of large-scale ordered domains of spheres in thin films of a narrowly dispersed diblock copolymer via thermodynamic control. Nanoscale. 2015;**7**:17756–17763. DOI: 10.1039/c5nr03837e

[59] Kimishima K, Saijo K, Koga T, Hashimoto T. Time-resolved high-resolution SAXS studies of OOT process and mechanism from hex-cylinder to BCC-sphere in a polystyrene-block-polyisoprene diblock copolymer. Macromolecules. 2013;**46**:9032–9044. DOI: 10.1021/ma401808p

[60] Ayoubi MA, Zhu K, Bo Nystrom B, Olsson U, Almdal K, Khokhlov AR, Piculell L. Morphological investigation of polydisperse asymmetric block copolymer systems of poly(styrene) and poly(methacrylic acid) in the strong segregation regime. Journal of Polymer Science, Part B: Polymer Physics. 2013;**51**:1657–1671. DOI: 10.1002/polb.23389

[61] Yoon J, Jung SY, Ahn B, Heo K, Jin S, Iyoda T, Yoshida H, Ree M. Order-order and order-disorder transitions in thin films of an amphiphilic liquid crystalline diblock copolymer. Journal of Physical Chemistry B. 2008;**112**:8486–8495. DOI: 10.1021/jp803664h

[62] Stein GE, Cochran EW, Katsov K, Fredrickson GH, Kramer EJ, Li X, Wang J. Symmetry breaking of in-plane order in confined copolymer mesophases. Physical Review Letters. 2007;**98**:158302. DOI: 10.1103/PhysRevLett.98.158302

[63] Huang YY, Hsu JY, Chen HL, Hashimoto T. Existence of fcc-packed spherical micelles in diblock copolymer melt. Macromolecules 2007;**40**:406–409. DOI: 10.1021/ma062149m

[64] Matsen MW. Cylinder-sphere epitaxial transitions in block copolymer melts. Journal of Chemical Physics. 2001;**114**:8165–8173. DOI: 10.1063/1.1365085

[65] Kimishima K, Koga T, Hashimoto T. Order–order phase transition between spherical and cylindrical microdomain structures of block copolymer. I. Mechanism of the transition. Macromolecules. 2000;**33**:968–977. DOI: 10.1021/ma991470k

Chapter 2

Amphiphilic Ionic Perylenediimides: Structures, Self-Assembly Studies and Biomedical Applications

Meizhen Yin and Baozhong Lü

Additional information is available at the end of the chapter

http://dx.doi.org/10.5772/intechopen.68399

Abstract

Amphiphilic ionic perylenediimides (AIPDIs) with well-defined structures have been widely studied, which involve abundant non-covalent interactions. Among these interactions, electrostatic interactions serve as the primary force that may be followed by other non-covalent interactions like π–π stacking. Taking advantage of these tunable interactions between simple AIPDI-building blocks, AIPDIs are widely used for constructing increasingly complex structures at varying scales. Besides, AIPDIs with outstanding photochemical stability exhibit high fluorescence quantum yields (FQYs) in aqueous solution, because hydrophilic substituents of AIPDIs can shield the inner perylene chromophores and weaken π–π stacking, contributing to the improvement of water solubility and the suppression of aggregation-caused quenching (ACQ). AIPDIs with excellent water solubility, strong FQYs and desired interactions with charged components in cells and tissues hold great promise for various biomedical applications, which can be concluded in three hierarchical levels, which is *in vitro,* live cell and tissue.

Keywords: amphiphilic ionic perylenediimides, fluorescence, self-assembly, biomedical applications

1. Introduction

Amphiphilic ionic self-assembly (AIS) is an important method to construct complex nanostructures by using non-covalent interactions of simple ionic molecular building blocks. Because of the well-defined structures, amphiphilic ionic molecules (AIMs) are studied widely in the field of AIS. Various nanostructures are obtained by the self-assembly of AIMs and the relationship between the morphologies and parameters was discovered by theoretical studies [1–8]. AIMs are designed and fabricated to form various nanostructures such as helical fibres, nanotubes, nanospheres and so on [9–12]. Among these works, Zhang and Eisenberg studied the effects of hydrophobic/hydrophilic ratio, chemical structure, temperature, concentration and solvent on the self-assembly behaviours of AIMs [13]. AIMs self-assemblies endow AIMs with enhanced stabilities, improved water solubility and increased fluorescence intensity, making them promising candidates in biomedical applications [10, 14–21]. As an attractive chromophore, perylene-3,4,9,10-tetracarboxylic acid diimides (PDIs) have been generally used in organic electronic and photovoltaic fields for their broad absorption range, high extinction coefficients, high fluorescence quantum yields (FQYs) as well as outstanding stability. In particular, amphiphilic ionic perylenediimides (AIPDIs) are found to form tunable supramolecular-ordering constructions by a well-known non-covalent force, that is, electrostatic interaction, which is operative in AIS. This electrophorus supramolecular process was firstly introduced by Faul and Antonietti [22], who found water-soluble molecular building blocks self-assembled with oppositely charged surfactants by electrostatic interactions as the primary force. This self-assembly approach has been successfully applied to AIPDIs for their hierarchical organization in aqueous solution and bulk solid state as well. AIPDIs exhibit good water solubility, strong fluorescence and desired interactions with charged components in cells and tissues. Moreover, AIPDIs emit fluorescence above 500 nm, which effectively avoid the interference of the autofluorescence from organism. All the advantages of AIPDIs make them excellent candidates for biomedical materials. In this chapter, we describe self-assembly behaviours of AIPDIs and focus on their chemical structures, self-assembly studies and biological applications.

2. Structures and self-assemblies of AIPDIs

In general, AIPDIs are synthesized by the modification with ionic substituents in the bay region, ortho or imide positions of PDIs (**Figure 1A**). The modifications in the imide positions of PDIs with ionic substituents will improve solubility and minimally affect the optical properties. By contrast, substituents in the bay region will lead to a twist of perylene core and changes in optical properties such as significant bathochromic shift. These modification strategies of

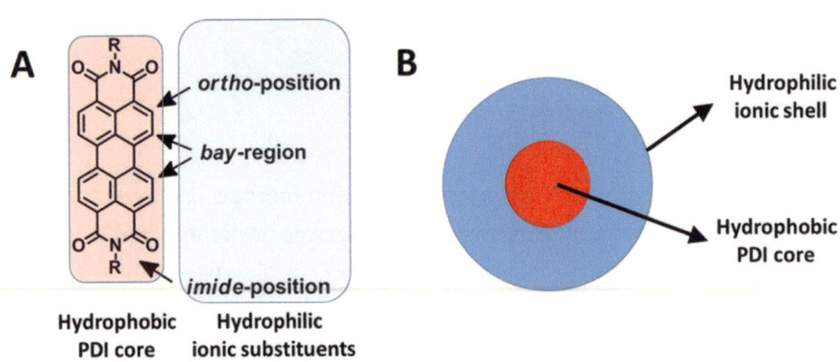

Figure 1. (A) General structure and hydrophobicity-hydrophilicity of AIPDIs, (B) schematic diagram of AIPDI core-shell structure.

PDIs in the imide position and bay region have attracted great interests in self-assembly behaviours because various intermolecular forces are introduced in this way. Besides, it will affect electronic and optical properties but not contort the two naphthalenes when selectively functionalized in the ortho positions of PDIs.

The fluorescence intensity and water solubility of modified PDIs are significantly improved compared with the unsubstituted PDIs [23]. Typically, the charged substituents can serve as the shell of PDI core, resulting in core-shell structure of AIPDIs (**Figure 1B**). In addition, the ionic substituents could provide electrostatic forces between the PDI molecules themselves or other charged guests. Intermolecular π-π stacking of PDIs could be attenuated by steric hindrance and electrostatic repulsion forces, leading to good water solubility and strong fluorescence of dye molecules.

2.1. Self-assembly of anionic AIPDIs

The self-assembly of AIPDI without surfactants was first studied by Ford in the 1980s [24]. In this fundamental research, concentration-dependent UV/vis absorption and fluorescence spectroscopy were studied. It showed that the anionic di(glycyl) PDI derivative **1a** (**Figure 2**) existed in an equilibrium between monomer and dimer of **1a** in alkaline aqueous solution with an association constant of $K = 1.0 \times 10^7\ M^{-1}$ at 24°C. The absorption maximum of dimers of this core unsubstituted dianionic AIPDI was strongly hypochromatic shift (32 nm) with considerable band broadening as compared with that for the monomer ($\lambda_{max} = 532$ nm, at $c = 8.7 \times 10^{-9}$ M), which was illuminated by molecular excitonic theory for parallel-oriented sandwich-type dimer. The fluorescence of the dimers was drastically quenched with a quantum yield of $\Phi_F < 0.02$ as compared with that of monomeric PDI anion. More interestingly, they found a considerable salt (NaCl) effect on the aggregation of dianionic AIPDI **1a**, which was discussed in light of counterion shielding of the electrostatic repulsion between the negatively charged AIPDI molecules. Although the synthesis and optical characterization of higher homologues of dicarboxylate **1a** with longer alkyl chains (**1b–d**) were reported already in the 1990s, their self-assembly behaviour has not been explored until today [25].

Malik and co-workers reported a pair of chiral AIPDIs **2** (**Figure 3**) modified with aspartic acid (L or D) [26]. The reddish-brown hydrogel with reddish fluorescence formed when the pH of **2** of aqueous solution decreased to 4. Atomic force microscopy (AFM) images revealed that dried

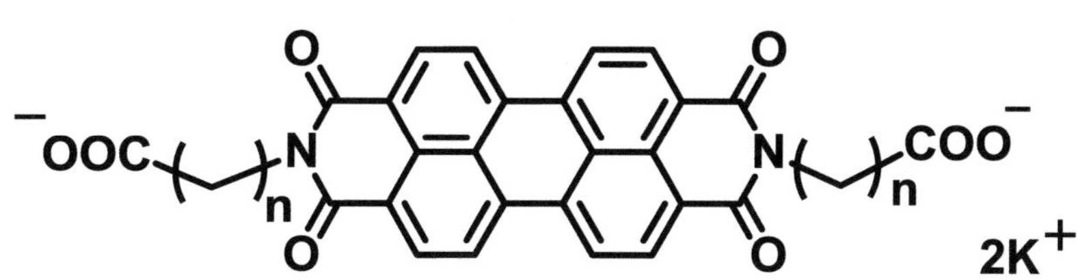

1a-d: n=1, 2, 4, 5

Figure 2. Anionic AIDPIs of **1a–d**.

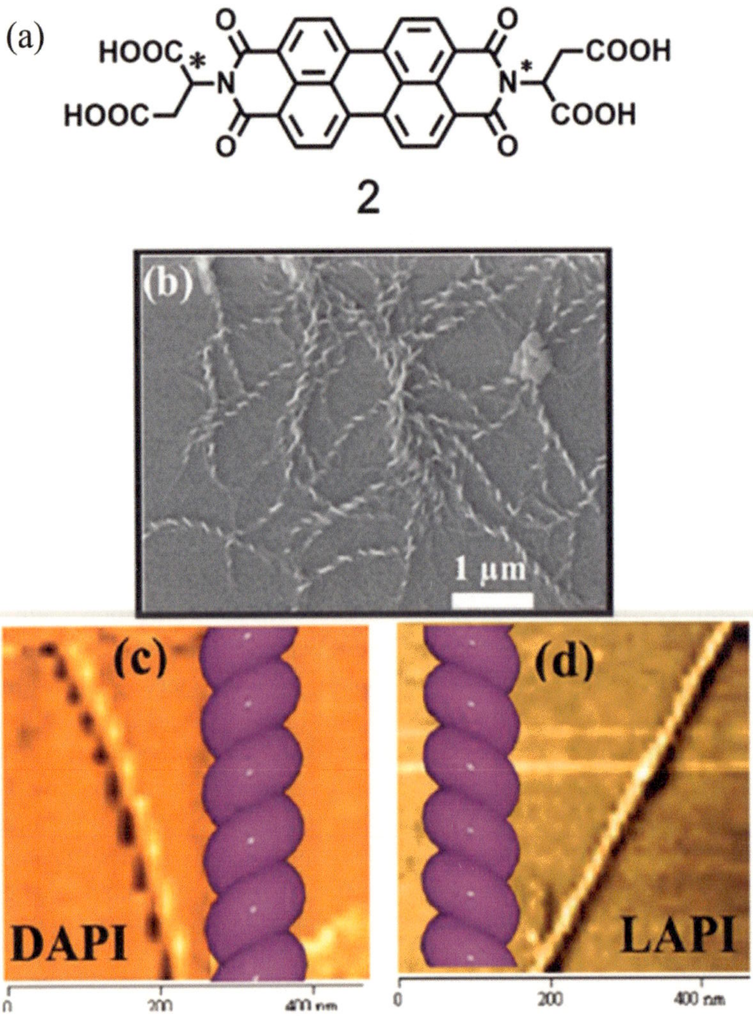

Figure 3. (a) Molecular structure of modified AIPDI **2**. (b) FE-SEM image of a **D-2**-dried gel. (c) AFM image of **D-2** gel and (d) **L-2** gel prepared on a mica substrate. Copyright (2013) The Royal Society of Chemistry.

L-2 or **D-2** gel formed helical fibres with a diameter of 20 nm and a pitch length of 20 nm. The perylene core played an important role during the formation of helical thick filaments. Further studies showed that the synergistic effect of π–π stacking, intermolecular hydrogen and chiral effect guided the aggregation and gelation in water during the decrease of pH.

2.2. Self-assembly of cationic AIPDIs

The synthesis and optical properties of AIPDIs **3a,b** modified with cationic ammonium side chains were reported already in the mid-1990s (**Figure 4**) [25, 27]. Subsequently, the self-assembly behaviour of AIPDIs **3a,b** has been reported. Trimethyl ammonium ethylene-functionalized dicationic AIPDI **3a** with iodide counterions has been studied by many

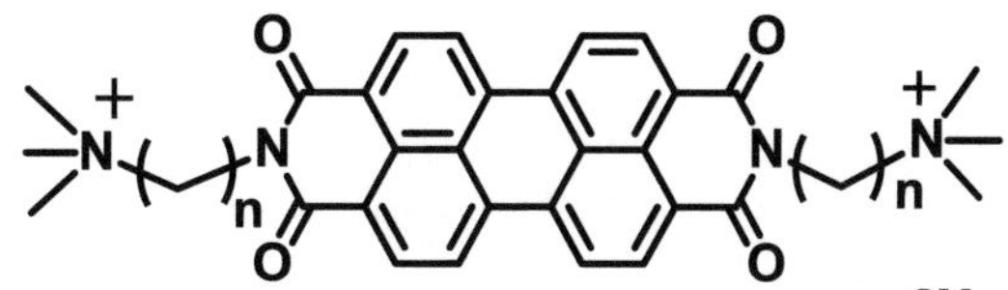

2X

3a: n=2 X=I$^-$
3b: n=3 X=I$^-$

Figure 4. Cationic AIPDIs **3**.

research groups with different objectives. Wei and co-workers first studied the self-assembly of AIPDI **3a** in the solvent of water and methanol [28]. They found that this cationic AIPDI **3a** could self-assemble into nanotubes with diameters of ca. 100–300 nm in aqueous medium, while long nanorods with diameters of ca. 200–300 nm were formed from methanol solution as verified by scanning electron microscopy (SEM) and transmission electron microscopy (TEM) (**Figure 5**). Then, Faul and co-workers reported the formation of supramolecular polymers by ionic self-assembly of the same cationic AIPDI **3a** with anionic copper-phthalocyanine tetrasulphonate in water. These supramolecular polymers are fibre-like structure as observed by different imaging techniques [29]. Very recently, this research group reported the self-assembly behaviours of a chiral derivative of **3a** containing a benzyl substituent at the chirality centre adjacent to the ammonium group. Their detailed studies showed that the self-assembly of this inherently chiral cationic AIPDI in water can be described according to the isodesmic model, and the molecular chirality of the imide side chains is transmitted to the self-assembled supramolecular structures due to induced CD effects that were observed [30]. AIPDI **3a** could also self-assemble together with other guests. Negatively charged tetrasulphonated zinc-porphyrin self-assembled with the AIPDI **3a** in water, resulting in a host-guest supramolecular assembly that showed photoinduced electron-transport property [31]. Hydrogel formed by co-assembling of this cationic AIPDI **3a** with a bicarboxylate functionalized oligo(phenylenevinylene) derivative [32]. Furthermore, layer-by-layer (LbL) deposition was used to proceed electrostatic self-assembly of this positively charged AIPDI **3a** together with two other derivatives. The deposited thin films incorporating less than 15 bilayers are composed of serpentine nanofibres as revealed by AFM [33]. Moreover, many complexes formed by self-assembly of ionic AIPDI **3a** with different negatively charged surfactants have thermotropic and lyotropic liquid crystalline (LC) properties. Trimethyl ammonium propylene-functionalized AIPDI **3b**, a higher homologue of **3a**, was reported to self-assemble with negatively charged adenosine triphosphate (ATP). Chiral superstructures with a right-handed helical stack of AIPDI **3b** formed via electrostatic forces and π–π interactions [34].

Yao and co-workers [35] synthesized a PID derivative with pyridyl in the bay region. They found that the self-assembly of amphiphilic AIPDI **4** (**Figure 6**) can be subtly controlled by tuning the environmental parameters. Mono-protonated dye **5** and doubly protonated dye **6** evoke new intermolecular interactions in the bay regions. These new secondary forces typically influence π-π stacking, thus leading to new aggregations. In particular, the ratio of

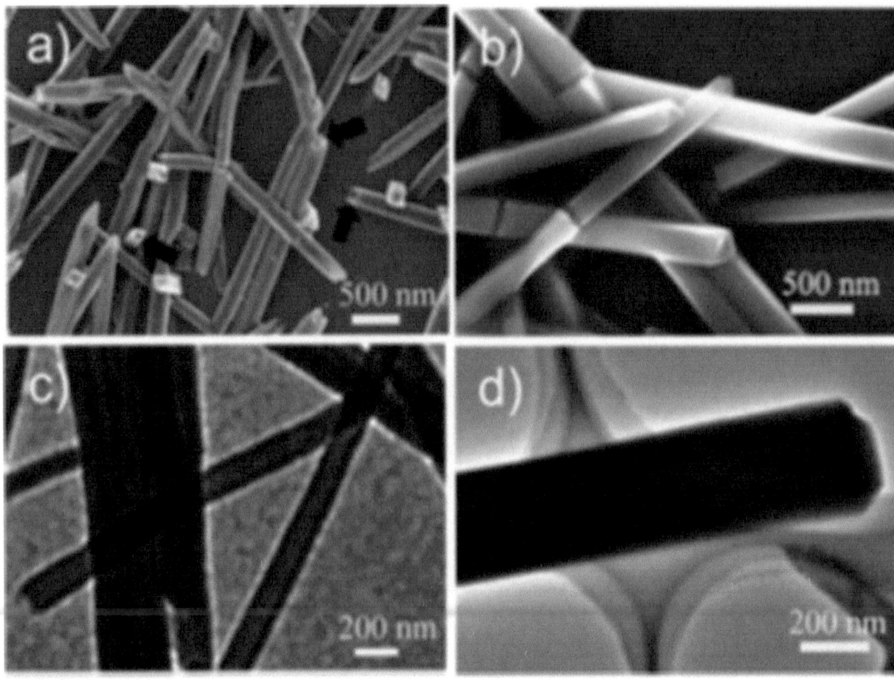

Figure 5. SEM (a, b) and TEM (c, d) images of nanotubes (a, c) and nanorods (b, d) formed from aqueous and methanol solutions (0.3 mM) of AIPDI **3a**, respectively. Copyright (2009) American Chemical Society.

hydrochloric acid (HCl)/methanol/water was utilized to adjust the solvent environment and self-assembly parameters, and then five different phases formed by subtle AIS (**Figure 7**). In this way, the representative self-assembled nanostructures were nanoparticles, nanotapes and corresponding one-dimensional (1D) aggregations, nanoplates, hollow nanospheres and corresponding 1D aggregations. As the most important factor, the water fraction (Rw)

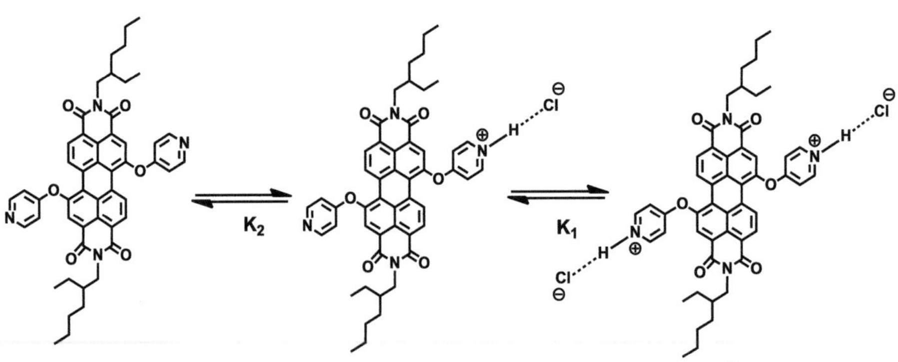

4 **5** **6**

Figure 6. Dynamic protonation among dyes **4–6**.

http://dx.doi.org/10.5772/intechopen.68399

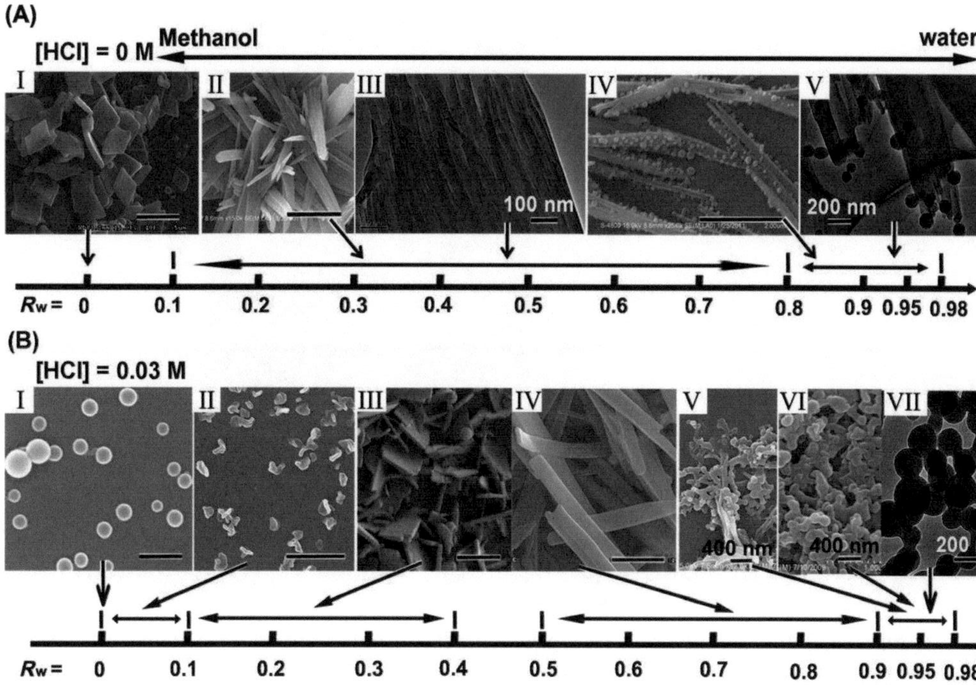

Figure 7. The typically 1D phase diagram and morphologies formed either from (A) chromophore **4** or (B) chromophore **6** equilibrating with dyes **5** and **4** at [HCl] = 0.03 M controlled by Rw. A-III, A-V and B-VII are TEM images. [dye] = 1 × 10^{-3}M is constant. The scale bar is 2 mm in the unlabelled pictures. Copyright (2012) John Wiley and Sons.

and the concentration of hydrochloric acid [HCl] play the crucial role to form the specific nanostructures. For example, nanotapes formed at low [HCl] and Rw values, whereas hollow nanospheres were observed when either the Rw is low or [HCl] is high, or both. To summarize, this work provides a phase diagram by tuning Rw and [HCl] as two variables. Such a self-assembly phase diagram summarizes the fine control on the self-assembled nanostructures, and inspires us to regulate the self-assembly of AIPDI by simply tuning the parameters.

2.3. Self-assembly of AIPDIs in charged surface

The attachment of charged dendritic substituents at the imide positions or in the bay regions of PDI can effectively suppress the aggregation of perylene chromophores and provide hydrophilia. The ionic nature of these dendrimers can self-assemble on the charged surface of gold or alumina template. Yin, Müllen et al. firstly reported a series of core-shell structures, by using shape-persistent polyphenylene dendrimer as a scaffold and a stabilizer. The outer flexible polymer shells contributed to proton-conducting property [36, 37]. Thus, a core-shell architecture **7** containing phosphonic acids was synthesized and its self-assembly on a modified gold substrate via electrostatic interaction was studied. AFM in **Figure 8** shows many uniformly dispersed globular particles, demonstrating unimolecular polymeric micelles [38].

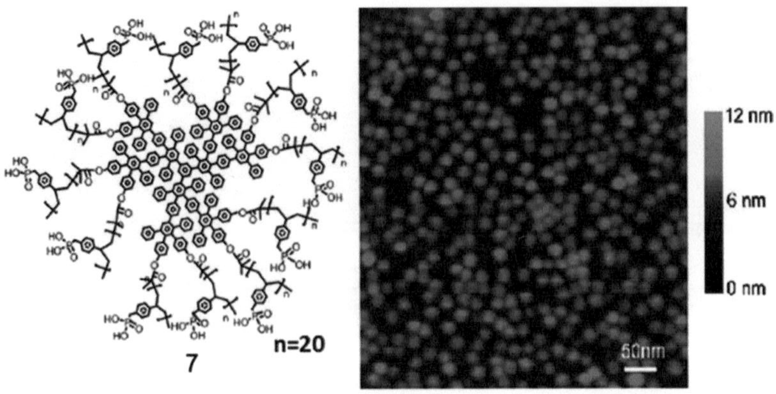

Figure 8. Core-shell structure 7 and AFM image of self-assembled unimolecular polymeric micelles of 7 on a modified gold film. Copyright (2012) John Wiley and Sons.

Recently, Yin, Müllen et al. reported a series of fluorescent core-shell AIPDI macromolecules based on polyphenylene dendrimers [39–41]. In order to introduce ionic functionalities such as amine and carboxylic acid, flexible polymer shell was attached to the shape-persistent cores. As representatives, core-shell AIPDIs **8c** and **9b** with opposite charges were synthesized (**Figure 9**) [42, 43]. AIPDIs **8c** and **9b** have a fluorescent PDI core attached with polyphenylene dendrimer and a flexible cationic or anionic polyelectrolytes shell, which contributes to the water solubility and pH sensitivity. Because of the polyelectrolyte nature, AIPDIs **8c** and **9b** could form pH-responsive globular unimolecular polymeric micelles in aqueous solution, which was confirmed by small-angle X-ray scattering (SAXS) [42]. The authors attributed the pH-responsive behaviour of AIPDIs **8c** and **9b** to the ionization or deionization of the star polyelectrolytes, which leads to the variation of volume phase and optical properties.

Templated self-assembly of the oppositely charged polymers was also investigated. LbL deposition of oppositely charged **8b** and **9a** in a porous alumina template was used to fabricate multilayer films. Due to the electrostatic interactions between charged **8b** and **9a** and alumina membrane, **8b** and **9a** assembled on the alumina surface during the LbL process. **Figure 10A** shows TEM image of self-assembled hollow nanotube after the removal of the

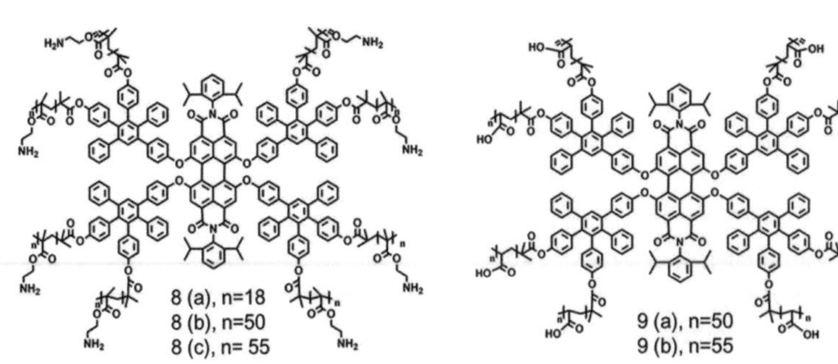

Figure 9. Core-shell of AIPDI macromolecules 8 and 9.

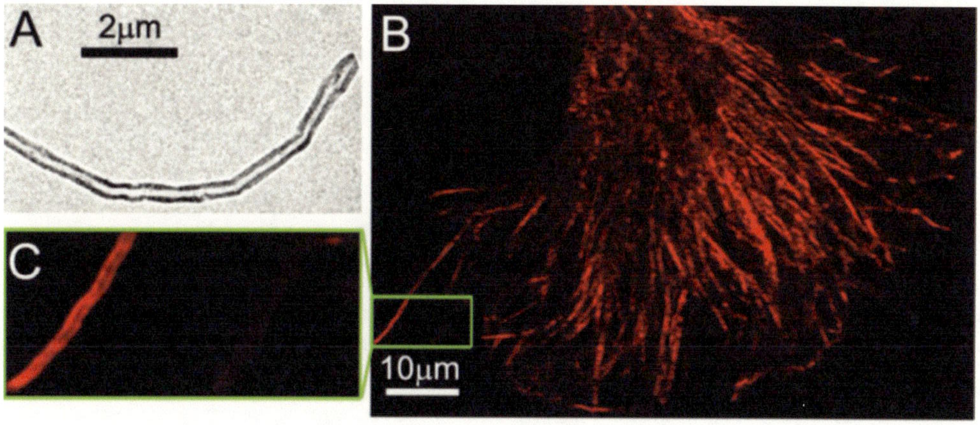

Figure 10. Self-assembled nanotubes fabricated by LbL deposition of AIPDI **8** and **9**. (A) TEM and (B, C) corresponding fluorescence microscope images of self-assembled nanotubes. Copyright (2011) John Wiley and Sons.

alumina template. Large-scale fluorescent nanotubes in solution were presented with a fluorescence microscope as shown in **Figure 10B** [43].

3. Biological applications of AIPDIs

Charged components are widely existed in organism, such as the negatively charged extracellular matrixes (ECMs), membranes, DNA and positively charged histone proteins. Therefore, water-soluble AIPDIs with charges on the surface would interact with the charged biological components through electrostatic forces. Moreover, AIPDIs have good water solubility and strong fluorescence, the fluorescence and absorption maxima of AIPDIs are above 500 nm, which could minimize the background noise from organism auto-fluorescence. All the advantages make them excellent candidates for biomedical materials. The biological applications of water-soluble AIPDIs are summarized below including imaging studies of living cells and tissues.

3.1. *In vitro* studies

Because of the fluorescence nature, the cationic AIPDIs were widely studied in the field of molecular detection in organism. As mentioned above, interactions between AIPDI and biological molecules are the most important part for researches in cells and tissues. Yin et al. have reported a type of fluorescent nanotubes fabricated by oppositely charged **10** and **11** which could serve as DNA biosensors (**Figure 11**) [43]. As shown in **Figure 12A**, positively charged AIPDI **10** on the outer surface of the nanotubes can interact with DNA by electrostatic forces. The complex of self-assembled nanotube and DNA showed obviously improved fluorescence intensity when compared with original nanotube because of the protection of PDI chromophore in aqueous solution. Subsequently, Yin et al. reported an ultrasensitive and selective DNA sensor by using the negatively charged AIPDI **12** as a donor and DNA targets labelled

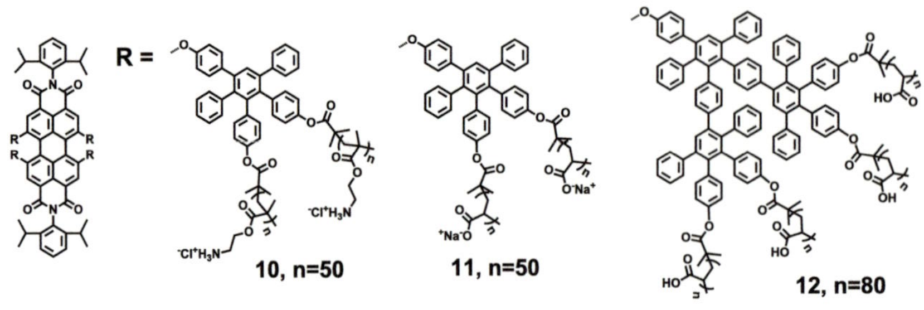

Figure 11. Chemical structures of AIPDIs 10–12.

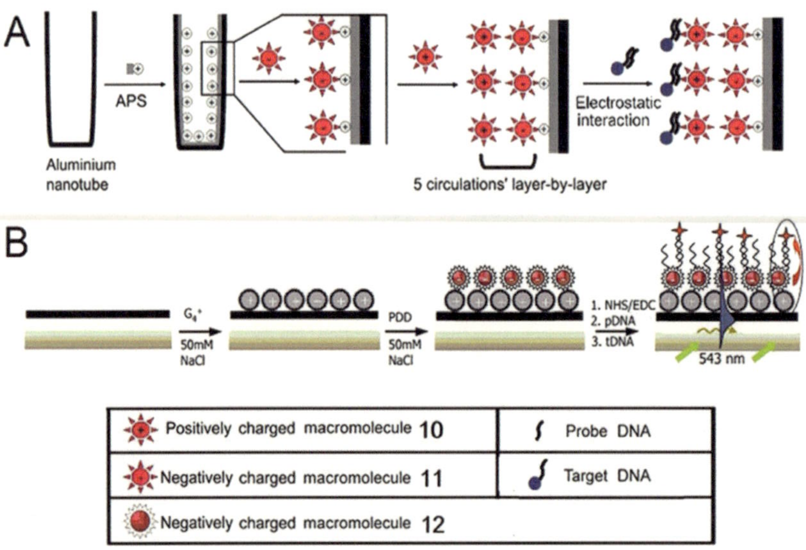

Figure 12. (A) Formation of fluorescent nanotubes by LbL deposition of **11** and **10**. Copyright 2011, John Wiley and Sons. (B) Energy transfer process between AIPDI **12** and Cy5. Copyright 2011, John Wiley and Sons.

by Cy5 as an acceptor (**Figure 12B**). Excitation energy was transferred from Cy5 to AIPDI **12** upon excitation at 543 nm. Due to energy transfer (ET), the sensor is ultrasensitive, selective and time-resolved (**Figure 12B**) [44].

Based on the interaction between probe **13** and nucleic acid aptamer, a fluorescent molecule **13** with switching properties is reported and used for sensitive and selective detection of proteins (**Figure 13**) [45]. In aqueous solution, probe **13** exists in monomeric forms, thus strong fluorescence can be observed (**Figure 13-1**). Nucleic acid aptamer is negatively charged, thus strong electrostatic interactions occur between AIPDI **13** and aptamer, leading to the aggregate formation. Due to the aggregation-caused quenching (ACQ) effect, the fluorescence intensity of the solution significantly decreases (**Figure 13-2**). Lysozyme has higher affinity to aptamer than that of **13**. As a result, aggregated **13** is replaced by the lysozyme and the intensity of fluorescence signal of **13** will be restored since **13** becomes monomeric state (**Figure 13-3**).

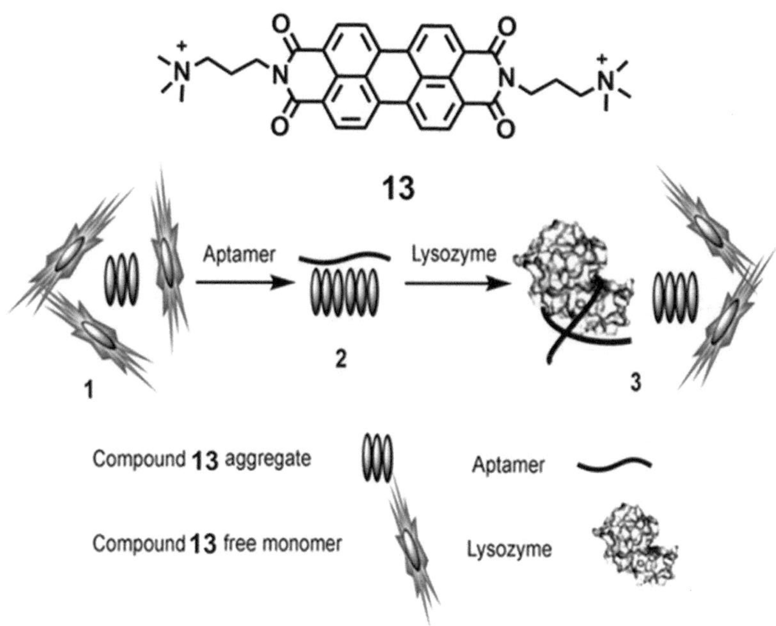

Figure 13. Chemical structures of cationic AIPDI **13** and schematic illustration for selective detection of lysozyme. Copyright (2010), John Wiley and Sons.

3.2. Live cell studies

With good selectivity and sensitivity, fluorescence bioimaging offers microscopic visualization of the organism. Thus, water-soluble AIPDIs with high fluorescence intensity have attracted great attention in bioimaging.

Yin et al. reported that dendritic AIPDIs **14a–c** and star-shaped polymers **15–16** are great gene delivery and bioimaging agents (**Figure 14**) [46, 47]. The fluorescence microscopy images indicate that all the cationic AIPDIs are internalized into cells and the cell viabilities of **14a–c**, **15** and **16** are all above 90%, suggesting that the AIPDIs have low cytotoxicity. All these AIPDIs molecules, **14a–c**, **15** and **16**, can serve as DNA delivery agents and can form probe/DNA complexes via electrostatic interaction with negatively charged DNA. The fluorescence images of **14c**/DNA and **16**/DNA complexes (**Figure 15**) showed that **14c** (A–A″) and **16** (B–B″) could deliver DNA into cells. Notably, the gene transfection efficiency of **14a–c** has an increased tendency with the increased generation of the dendritic molecules due to more positive charges and the larger external branches of dendritic molecules.

3.3. Tissue imaging

Fluorescence labelling, the process of attaching a reactive probe to a functional group of target molecule with covalent or non-covalent interactions, is usually used for tissue imaging. Generally, the target structure, such as antibodies, proteins, amino acids and peptides, is charged. Thus, based on the electrostatic interactions with those charged electrolytes, AIPDIs have unique advantages in tissue imaging.

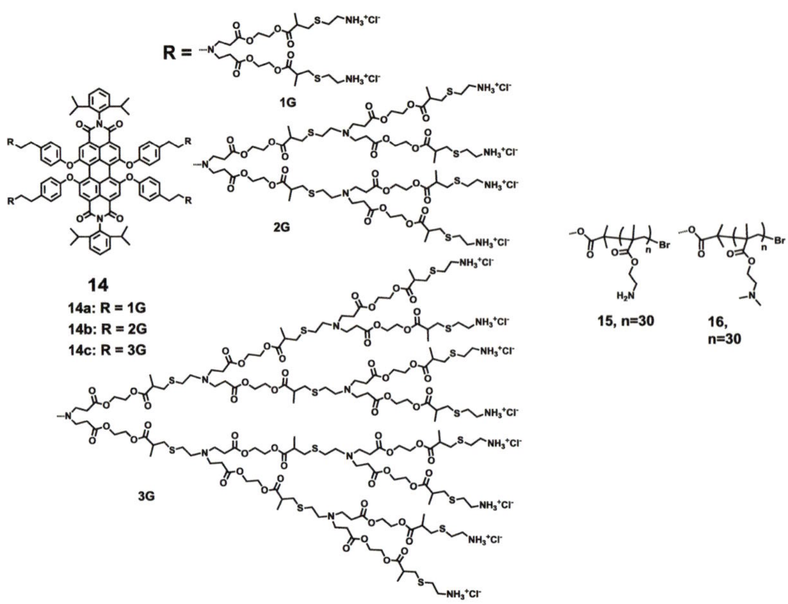

Figure 14. Chemical structures of macromolecules **14–16**.

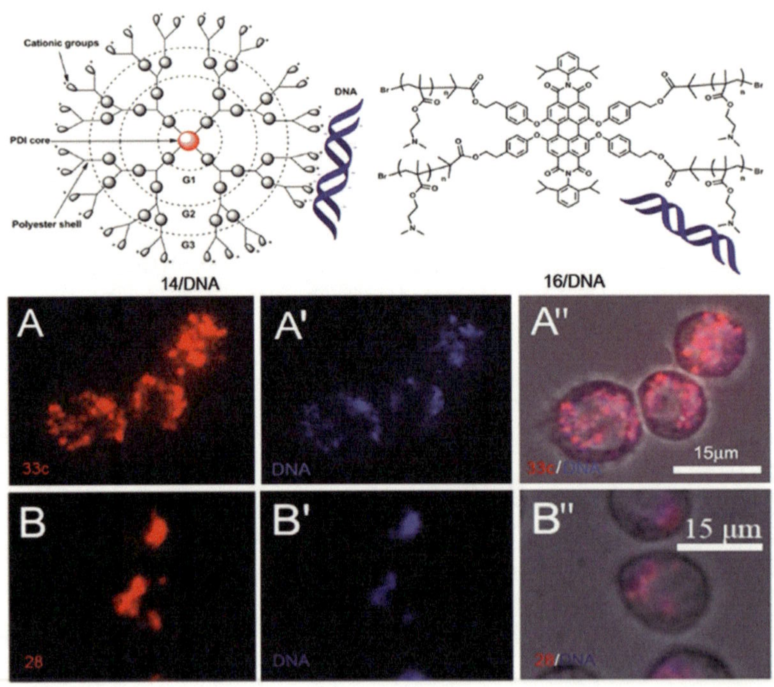

Figure 15. Schematic illustration of the complexes formed by **14c** or **16** and DNA. Fluorescence images of the **14c** (A) and **16** (B) internalizing into cells; A′ and B′ are fluorescence images of DNA labelled by CXR Reference Dye; A″ and B″ are images of the complex of **14c**/DNA and **16**/DNA. Copyright (2013), Royal Society of Chemistry for Ref. [47] and copyright (2014), American Chemical Society for Ref. [46].

http://dx.doi.org/10.5772/intechopen.68399

The star polymer **17** with negative charges cannot enter the living cells [48], but can specifically label cell nucleus in fixed tissues via binding to positively charged nuclear proteins [49]. As shown in **Figure 16**, the red staining of AIPDI **17** shows almost no overlap with the membrane marker CD8-GF (green part, **Figure 16A**), or cytoskeletal marker anti-α-tubulin antibodies (green, **Figure 16B**). In **Figure 16C** and **D**, the staining of AIPDI **17** has the same pattern with DAPI (commercial DNA dye, blue, **Figure 16C″**) and demonstrates a prominently co-localization effect with histone H4 (green, **Figure 16D″**AIPDI **17**, bearing multiple carboxyl groups, could specifically bind positively charged histone, leading to specific labelling of cell nucleus. This work provides a fascinating substitute for conventional costly antibodies or chromophores with broad emission such as commercial dye, DAPI.

By contrast, the positively charged star polymer **18** could enter the live cells and serve as label agent [46]. Yin and co-workers [50] firstly reported an exciting application of **18** in

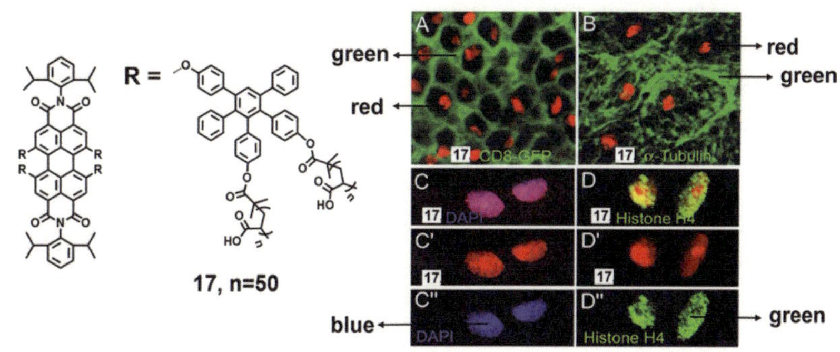

Figure 16. Chemical structure of **17** and Drosophila larval tissues observed by confocal microscopy. (A, green) Double staining of **17** (red) with CD8-GFP, (B, green) anti-α-tubulin. DAPI (blue, C-C″), and histone H4 (green, D-D″). C and D are combined channels while C′, C″ and D′, D″ are single channels. Copyright (2008), John Wiley and Sons.

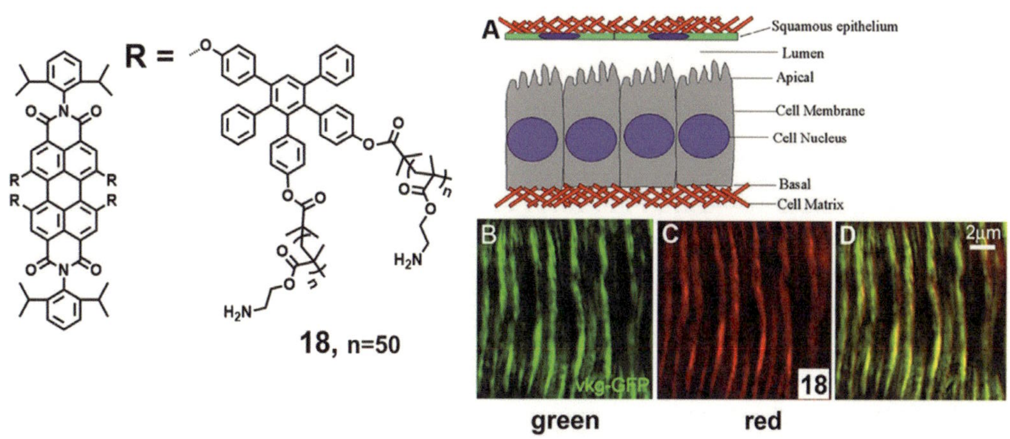

Figure 17. Chemical structure of **18** and illustration of extracellular matrix in a small central section of the Drosophila wing disc epithelium (A). ECM staining with Vkg-GFP (B, green) and **18b** (C, red) in living tracheal epithelium and merged channels (D). Copyright (2008), American Chemical Society.

2008, which could be used to directly mark the extracellular matrix network (**Figure 17A**) and visualize ECM network. The positively charged **18** could strongly bind with the negatively charged ECM components via electrostatic interactions in fixed or living tissue. **Figure 17** shows the high-resolution confocal microscopy image of ECM stained with **18** (red, **Figure 17C**) in living trachea epithelium. **18b** has the same specificity and much higher labelling speed compared with Vkg-GFP (green, **Figure 17D**). Besides, the fluorescence intensity of **18** increased remarkably after binding ECM, indicating the direct interaction between **18** and negatively charged ECM components. All the above results revealed that **18** could serve as ECM-labelling agent with high efficiency and specificity at the same time.

4. Conclusions and perspectives

This chapter gives a brief summary of the structures, self-assembly and biology applications of AIPDIs. AIPDIs have emerged as promising candidates for AIS and used in bio-applications. The synthetic strategies of AIPDIs are composed of the introduction of ionic substituents in the bay region, ortho or imide positions of PDIs. AIPDIs were found to form supramolecular-ordering construction by electrostatic interaction and other synergistic non-covalent interactions. By adjusting self-assembly conditions, AIPDIs have been successfully applied to form hierarchical organization in aqueous solution and bulk solid state. The biological applications involve studies *in vitro*, live cell and tissue imaging. Electrostatic interactions between AIPDIs and charged components in cells and tissues play an important role in these biological applications. To conclude, due to the high stability, water solubility, low cytotoxicity and desirable self-assembly behaviours, AIPDIs have great potential in biological applications and are still a topic of uninterrupted study.

Acknowledgements

This work was financially supported by the National Natural Science Foundation of China (21574009 and 51521062), Beijing Collaborative Innovative Research Center for Cardiovascular Diseases, and the Higher Education and High-quality and World-Class Universities (PY201605).

Author details

Meizhen Yin* and Baozhong Lü

*Address all correspondence to: yinmz@mail.buct.edu.cn

State Key Laboratory of Chemical Resource Engineering, Beijing Laboratory of Biomedical Materials, Beijing University of Chemical Technology, Beijing, China

References

[1] Motala-Timol S, Jhurry D, Zhou J, Bhaw-Luximon A, Mohun G, Ritter H. Amphiphilic poly(l-lysine-b-caprolactone) block copolymers: Synthesis, characterization, and solution properties. Macromolecules. 2008;**41**:5571-5576

[2] Chan S-C, Kuo S-W, Lu C-H, Lee H-F, Chang F-C. Syntheses and characterizations of the multiple morphologies formed by the self-assembly of the semicrystalline P4VP-b-PCL Diblock copolymers. Polymer. 2007;**48**:5059-5068

[3] Tung PH, Kuo SW, Chan SC, Hsu CH, Wang CF, Chang FC. Micellization and the surface hydrophobicity of amphiphilic poly(vinylphenol)-block-polystyrene block copolymers. Macromolecular Chemistry and Physics. 2007;**208**:1823-1831

[4] Colombani O, Ruppel M, Schubert F, Zettl H, Pergushov DV, Müller AH. Synthesis of poly(n-butyl acrylate)-block-poly(acrylic acid) diblock copolymers by ATRP and their micellization in water. Macromolecules. 2007;**40**:4338-4350

[5] Colombani O, Ruppel M, Burkhardt M, Drechsler M, Schumacher M, Gradzielski M, Schweins R, Müller AH. Structure of micelles of poly(n-butyl acrylate)-block-poly(acrylic acid) diblock copolymers in aqueous solution. Macromolecules. 2007;**40**:4351-4362

[6] Bhargava P, Tu Y, Zheng JX, Xiong H, Quirk RP, Cheng SZ. Temperature-induced reversible morphological changes of polystyrene-block-poly(ethylene oxide) micelles in solution. Journal of the. American Chemical Society. 2007;**129**:1113-1121

[7] Liu L, Gao X, Cong Y, Li B, Han Y. Multiple morphologies and their transformation of a polystyrene-block-poly(4-vinylpyridine) block copolymer. Macromolecular Rapid Communications. 2006;**27**:260-265

[8] Gohy J-F, Creutz S, Garcia M, Mahltig B, Stamm M, Jérôme R. Aggregates formed by amphoteric diblock copolymers in water. Macromolecules. 2000;**33**:6378-6387

[9] Du J, Chen Y. Organic-inorganic hybrid nanoparticles with a complex hollow structure. Angewendte Chemie International Edition. 2004;**43**:5084-5087

[10] Du J, Tang Y, Lewis AL, Armes SP. pH-sensitive vesicles based on a biocompatible zwitterionic diblock copolymer. Journal of the American Chemical Society. 2005; **127**:17982-17983

[11] Blanazs A, Madsen J, Battaglia G, Ryan AJ, Armes SP. Mechanistic insights for block copolymer morphologies: How do worms form vesicles? Journal of the American Chemical Society. 2011;**133**:16581-16587

[12] Yin M, Bauer R, Klapper M, Müllen K. Amphiphilic multicore-shell particles based on polyphenylene dendrimers. Macromolecular Chemistry and Physics. 2007;**208**:1646-1656

[13] Yin M, Cheng Y, Liu M, Gutmann JS, Müllen K. Nanostructured TiO_2 films templated by amphiphilic dendritic core-double-shell macromolecules: From isolated nanorings to continuous 2D mesoporous networks. Angewendte Chemie International Edition. 2008;**120**:8528-8531

[14] Discher DE, Eisenberg A. Polymer vesicles. Science. 2002;**297**:967-973

[15] Chen W, Meng F, Cheng R, Zhong Z. pH-Sensitive degradable polymersomes for triggered release of anticancer drugs: A comparative study with micelles. Journal of Controlled Release. 2010;**142**:40-46

[16] Yan Q, Zhou R, Fu C, Zhang H, Yin Y, Yuan J. CO_2-responsive polymeric vesicles that breathe. Angewendte Chemie International Edition. 2011;**123**:5025-5029

[17] Szostak JW, Bartel DP, Luisi PL. Synthesizing life. Nature. 2001;**409**:387-390

[18] Roodbeen R, van Hest J. Synthetic cells and organelles: Compartmentalization strategies. Bioessays. 2009;**31**:1299-1308

[19] Alpermann T, Rüdel K, Rüger R, Steiniger F, Nietzsche S, Filiz V, Förster S, Fahr A, Weigand W. Polymersomes containing iron sulfide (FeS) as primordial cell model: For the investigation of energy providing redox reactions. Origins of Life and Evolution of Biospheres. 2011;**41**:103-119

[20] Poulos TL. The Janus nature of heme. Natural Product Reports. 2007;**24**:504-510

[21] Wan X, Liu T, Liu S. Synthesis of amphiphilic tadpole-shaped linear-cyclic diblock copolymers via ring-opening polymerization directly initiating from cyclic precursors and their application as drug nanocarriers. Biomacromolecules. 2011;**12**:1146-1154

[22] Faul CFJ, Antonietti M. Ionic self-assembly: Facile synthesis of supramolecular materials. Advanced Materials. 2003;**15**:673-683

[23] Nakazono S, Imazaki Y, Yoo H, Yang J, Sasamori T, Tokitoh N, Cedric T, Kageyama H, Kim D, Shinokubo H. Regioselective Ru-catalyzed direct 2,5,8,11-alkylation of perylene bisimides. Chemistry—European Journal. 2009;**15**:7530-7533

[24] Ford WE. Photochemistry of 3,4,9,10-perylenetetracaboxylic dianhydride dyes: Visible absorption and fluorescence of di(glycyl)- imide derivative monomer and dimer in basic aqueous solutions. Journal of Photochemistry. 1987;**37**:189-204

[25] Schnurpfeil G. Syntheses of uncharged, positively and negatively charged 3,4,9,10-perylene-bis(dicarboximides). Dyes Pigmentation. 1995;**27**:339-350

[26] Sukul PK, Singh PK, Maji SK, Malik S. Aggregation induced chirality in a self-assembled perylene based hydrogel: Application of the intracellular pH measurement. Journal of Materials Chemistry B. 2013;**1**:153-156

[27] Deligeorgiev T, Zaneva D, Petkov I, Timcheva I, Sabnis R. Synthesis and properties of fluorescent bis-quaternized perylene dyes. Dyes Pigmentation. 1994;**24**:75-81

[28] Huang Y, Quan B, Wei Z, Liu G, Sun L. Self-assembled organic functional nanotubes and nanorods and their sensory properties. Journal of Physics and Chemistry C. 2009;**113**:3929-3933

[29] Guan Y, Yu S, Antonietti M, Bottcher C, Faul C. Synthesis of supramolecular polymers by ionic self-assembly of oppositely charged dyes. Chemistry—A European Journal. 2005;**11**:1305-1311

[30] Echue G, Lloyd G, Faul C. Chiral perylene diimides: Building blocks for ionic self-assembly. Chemistry—A European Journal. 2015;**21**:5118-5128

[31] Supur M, Fukuzumi S. Photodriven electron transport within the columnar perylenediimide nanostructures self-assembled with sulfonated porphyrins in water. Journal of Physics and Chemistry C. 2012;**116**:23274-23282

[32] Rao K, George S. Supramolecular alternate Co-assembly through a non-covalent amphiphilic design: Conducting nanotubes with a mixed D-A structure. Chemistry—A European Journal. 2012;**18**:14286-14291

[33] Weitzel C, Everett T, Higgins D. Aggregation and its influence on macroscopic in-plane organization in thin films of electrostatically self-assembled perylene-diimide/polyelectrolyte nanofibers. Langmuir. 2009;**25**:1188-1195

[34] Ma T, Li C, Shi G. Optically active supramolecular complex formed by ionic self-assembly of cationic perylenediimide derivative and adenosine triphosphate. Langmuir. 2008;**24**:43-48

[35] Zhang Z, Zhan C, Zhang X, Zhang, S, Huang J, Li A, Yao J. A self-assembly phase diagram from amphiphilic perylene diimides. Chemistry—A European Journal. 2012;**18**, 12305-12313

[36] Kainthan RK, Brooks DE. Unimolecular micelles based on hydrophobically derivatized hyperbranched polyglycerols: Biodistribution studies. Bioconjugate Chemistry. 2008;**19**:2231-2238

[37] He B, Chu Y, Yin M, Müllen K, An C, Shen J. Fluorescent nanoparticle delivered dsRNA toward genetic control of insect pests. Advanced Materials. 2013;**25**:4580-4584

[38] Yin M, Kang N, Cui G, Liu Z, Wang F, Yang W, Klapper M, Müllen K. Synthesis, electrochemical properties and self-assembly of a proton-conducting core-shell macromolecule. Chemistry—A European Journal. 2012;**18**:2239-2243

[39] Xu Z, He B, Wei W, Liu K, Yin M, Yang W, Shen J. Highly water-soluble perylenediimide-cored poly(amido amine) vector for efficient gene transfection. Journal of Materials Chemistry B. 2014;**2**:3079-3086

[40] Morgenroth F, Reuther E, Müllen K. Polyphenylen-dendrimere: von dreidimensionalen zu zweidimensionalen strukturen. Angewendte Chemie International Edition in England. 1997;**36**:631-634

[41] Liu X, He B, Xu Z, Yin M, Yang W, Zhang H, Cao J, Shen J. A functionalized fluorescent dendrimer as a pesticide nanocarrier: Application in pest control. Nanoscale. 2015;**7**:445-449

[42] You S, Cai Q, Müllen K, Yang W, Yin M. pH-sensitive unimolecular fluorescent polymeric micelles: From volume phase transition to optical response. Chemical Communication. 2014;**50**:823-825

[43] Yin M, Feng C, Shen J, Yu Y, Xu Z, Yang W, Knoll W, Müllen K. Dual-responsive interaction to detect DNA on template-based fluorescent nanotubes. Small. 2011;**7**:1629-1634

[44] Feng C, Yin M, Zhang D, Zhu S, Caminade AM, Majoral JP, Müllen K. Fluorescent core-shell star polymers based bioassays for ultrasensitive DNA detection by surface plasmon fluorescence spectroscopy. Macromolecular Rapid Communications. 2011;**32**:679-683

[45] Wang B, Yu C. Fluorescence turn-on detection of a protein through the reduced aggregation of a perylene probe. Angewendte Chemie International Edition. 2010;**49**:1485-1488

[46] You S, Cai Q, Zheng Y, He B, Shen J, Yang W, Yin M. Perylene-cored star-shaped polycations for fluorescent gene vectors and bioimaging. ACS Applied Materials Interfaces. 2014;**6**:16327-16334

[47] Xu Z, He B, Shen J, Yang W, Yin M. Fluorescent water-soluble perylenediimide-cored cationic dendrimers: Synthesis, optical properties, and cell uptake. Chemical Communications. 2013;**49**:3646-3648

[48] Yin M, Kuhlmann CR, Sorokina K, Li C, Mihov G, Pietrowski E, Koynov K, Klapper M, Luhmann HJ, Weil T. Novel fluorescent core–shell nanocontainers for cell membrane transport. Biomacromolecules. 2008;**9**:1381-1389

[49] Yin M, Shen J, Gropeanu R, Pflugfelder GO, Weil T, Müllen K. Fluorescent core/shell nanoparticles for specific cell-nucleus staining. Small. 2008;**4**:894-898

[50] Yin M, Shen J, Pflugfelder GO, Müllen K. A fluorescent core-shell dendritic macromolecule specifically stains the extracellular matrix. Journal of the American Chemical Society. 2008;**130**:7806-7807

Chapter 3

Supramolecular Materials Based on Ionic Self-Assembly: Structure, Property, and Application

Jinglin Shen, Shiling Yuan and Xia Xin

Additional information is available at the end of the chapter

http://dx.doi.org/10.5772/67906

Abstract

The technique of ionic self-assembly (ISA), on the basis of electrostatic interactions, is a powerful tool to create new material nanostructures and chemical objects due to its advantages of facility, reliability, cost saving, flexibility, and universality. It has attracted great attention because of its promising applications in catalysis, drug delivery, and molecular detection. This review focuses on recent advances in the construction of self-assemblies with different morphologies on the basis of ISA strategy and its applications. The ISA method provides an opportunity to generate complex and hierarchical assemblies with tunable properties, which is regarded as a very promising case of supramolecular chemistry.

Keywords: ionic self-assembly, amphiphilic molecule, electrostatic interaction, supramolecular materials

1. Introduction

Supramolecular self-assembly, which makes use of molecules instead of atomic units, offers a bottom-up approach to the construction of new materials on multiple length scales without complex organic synthesis [1–10]. The driving forces for supramolecular chemistry are noncovalent interaction ranging from host-guest interaction [11], hydrogen bonding [12], van der Waals interaction [13], π—π stacking [14], electrostatic interactions [15] and hydrophobic effect [16]. **Table 1** summarizes most of this noncovalent interaction, as well as of its structure-determining properties [17]. Among them, the theme of ionic self-assembly (ISA) on the basis of electrostatic interactions was first described by Faul and Antonietti [17]. Different with the simple coulombic binding of salts, the ISA is usually accompanied by a cooperative binding (π—π stacking, hydrophobic effect, and van der Waals interaction) which propagates toward the final self-assembly structures [17]. The ISA strategy is very easily available, reliable, and flexible and is much broader in application than multiple hydrogen bonding or stable metal coordination.

In this chapter, we provide an overview of work that contributed to establishing functional soft materials on the basis of ISA strategy and advancing the useness in the several areas such as soft matter template, luminescent materials, and inorganic-organic hybrid materials.

Type of interaction	Strength [kJ mol^{-1}]	Range	Character
Van der Waals	51	Short	Non-selective, non-directional
H-bonding	5–65	Short	Selective, directional
Coordination binding	50–200	Short	Directional
"Fit interaction"	10–100	Short	Very selective
"Amphiphilic"	5–50	Short	Non-selective
Ionic	50–250[a]	Long	Non-selective
Covalent	350	Short	Irreversible

[a]Dependent on solvent and ion solution; data are for organic media.

Table 1. Methods of self-assembly listed by invoked secondary interactions [17].

2. Dye-surfactant ISA materials

Dye molecules are almost the ideal building blocks for supramolecular chemistry, because they are easily available and have multiple functional groups, which possess a defined and regular shape (extended π system) and can facilitate mutual interactions, inducing them to have photoelectric functions and can form more complex structures, such as soft gels, liquid crystals, as well as plate-like and needle-like single crystals [18, 19]. Besides, the surfactant molecule plays a special role in surface chemistry. Especially, surface active ionic liquids (SAILs) such as imidazolium-, pyrrole-based ionic liquids have attracted widespread interest for their outstanding performance in biomaterials, photoelectricity, and environmentally friendly trades, since they possess good biocompatibility, low critical micellar concentrations, wide liquid range, and high thermal stability [20, 21]. Thus, the introduction of charged dye molecules or surfactants as ISA components will generate novel properties and structures of the systems and enrich the ranges and contents of ISA.

However, structures of supramolecular materials constructed by dye-surfactant have been limited to 1D and 2D structures, such as fiber and flake morphologies, and mostly have no fluorescent property, which greatly restricts the application of the material [22, 23]. For example, highly ordered fibers constructed by dyes and surfactants have been reported by Faul [24]. Zheng et al. also prepared highly ordered supramolecular microfibers through ISA strategy from complexes of the SAIL N-tetradecyl-N-methylpyrrolidinium bromide (C_{14}MPB) and anionic dye methyl orange (MO) with the aiding of patent blue VF sodium salt (PB) (**Figure 1a–g**) [23]. The crystal structure of microfibers can be indexed as a typical hexagonal columnar mesophase by XRD (**Figure 1h**). The driving force for the formation of microfiber is hydrogen-bonding interactions, π–π stacking interactions, especially, the role of PB whose big disc-like structure favors the formation of a one-dimensional supramolecular material.

Based on the literature work, Xin and coworkers constructed giant vesicles (1–10 μm) via a facile ISA strategy using an anionic dye Acid Orange II (AO) and an oppositely charged ionic-liquid-type cationic surfactant 1-tetradecyl-3-methylimidazolium bromide (C_{14}mimBr) (**Figure 2a–c**) [25]. It is concluded that the electrostatic interaction, hydrophobic effect, and π–π stacking interaction play key roles in this self-assembly process. Importantly, the giant vesicles can act as a smart microcarrier to load and release carbon quantum dots (CQDs) under control (**Figure 2d, e**). Besides, the giant vesicles could also be applied as a microrector to synthesize monodispersed Ag nanoparticles with diameter of about 5–10 nm which exhibited the ability to catalyze reduction of 4-nitroaniline (**Figure 2f, g**). Therefore, it is indicated that the AO/C_{14}mimBr assemblies hold promising applications in the areas of microencapsulation, catalyst support, and light weight composites owing to their huge sizes, and large microcavities.

Moreover, giant vesicle could also be prepared by anionic dye MO and an oppositely charged SAILs C_{14}mimBr (**Figure 3a–d**) [26]. The giant vesicle performs fluorescence property owing to the break of π–π stacking of MO molecules (**Figure 3c**), and the formation of giant vesicle was confirmed by the fusion of small vesicle with the trace of TEM and DLS (**Figure 3d, i**). The

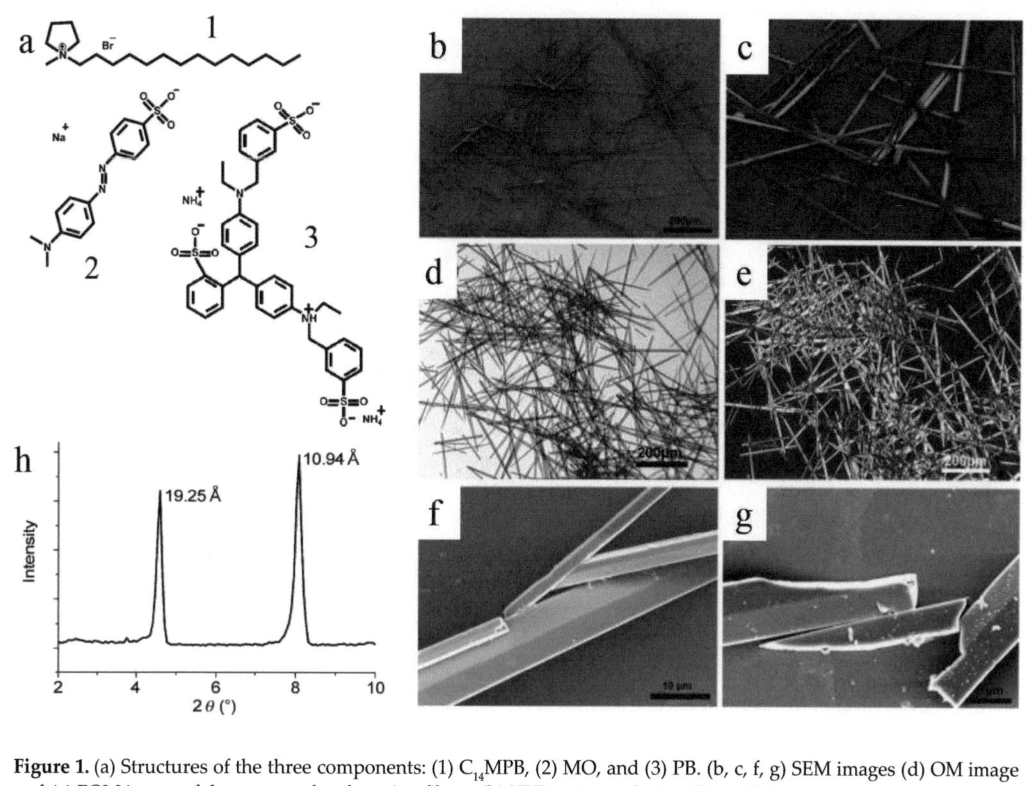

Figure 1. (a) Structures of the three components: (1) C_{14}MPB, (2) MO, and (3) PB. (b, c, f, g) SEM images (d) OM image and (e) POM image of the supramolecular microfibers. (h) XRD pattern of microfibers [23].

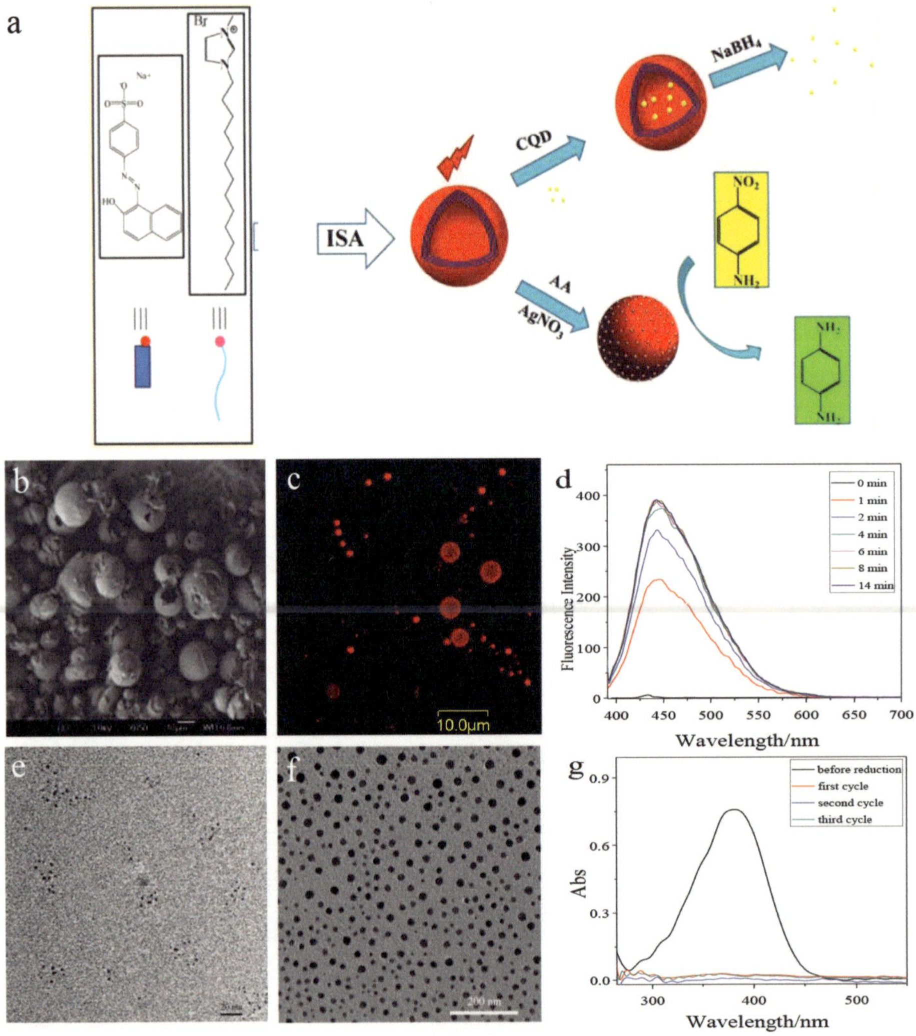

Figure 2. (a) Schematic demonstration of formation of giant vesicle and its applications for smart microcarrier and microreactor. (b) SEM and (c) CLSM images of giant vesicles formed by 0.5 mmol L^{-1} AO/0.5 mmol $L^{-1}C_{14}$mimBr. (d) Releasing of CQDs with the addition of $NaBH_4$. (e) TEM image of CQDs. (f) TEM image of Ag nanoparticle. (g) UV-vis spectra of 4-nitroaniline before and after reduction [25].

MO/C_{14}mimBr system also shows an interesting pH chromism phenomenon. Consequently, the MO/C_{14}mimBr system displays abundant aggregate morphologies with the changing of pH that complex fluorescent structures formed at pH > 4 and simple nonfluorescent structures formed at pH < 4 (**Figure 3e–h**). This is attributed to that MO moieties take a more conjugated coplanar state bridged by N=N under alkaline conditions, whereas MO takes a quinoid structure crossed by N–N resulting the break of conjugated configuration with the decreasing of pH (**Figure 3j, k**) [27]. Thus, the fluorescence behavior can be predicted with the color change directly visible to the naked eye by changing the pH. It is expected that the facile and innovative design of supramolecular material by the ISA strategy could be used as pH detection probes and microreactors.

http://dx.doi.org/10.5772/67906

Recently, one kind of biomolecule, bile salts, has been developed as building blocks for various functional nanomaterials [28, 29]. All of the bile salts possess a rigid, nearly planar hydrophobic steroid backbone and have polar hydroxyl groups on the concave α-face and methyl groups on the convex β-face [30, 31]. The ionic head with a carboxyl group is linked to the steroid ring through a short alkyl chain. Sodium deoxycholate (NaDC) is a bile salt and also an important biological surfactant which widely exists in the body of vertebrate [32]. The interesting structure of NaDC leads to novel and abundant self-assembly behavior in solution. For example, Xin and coworkers chose NaDC and a cationic dye (RhB) to construct soft materials [33]. In this system, different morphologies with high hierarchies can be reversibly controlled by varying the ratio of the two components (NaDC and RhB) and that the morphologies can switch between porous microspheres and urchin-like structures (**Figure 4a–c**). The robust hierarchical nanostructure performs superhydrophobicity (water contact angle reach to 137.1° for porous microspheres and 134.2° for urchin-like structures) (**Figure 4a, c**), which can be used to fabricate an anti-wetting surface. More interestingly, the interaction between NaDC and RhB can restrict the intramolecular motion and intramolecular charge transfer

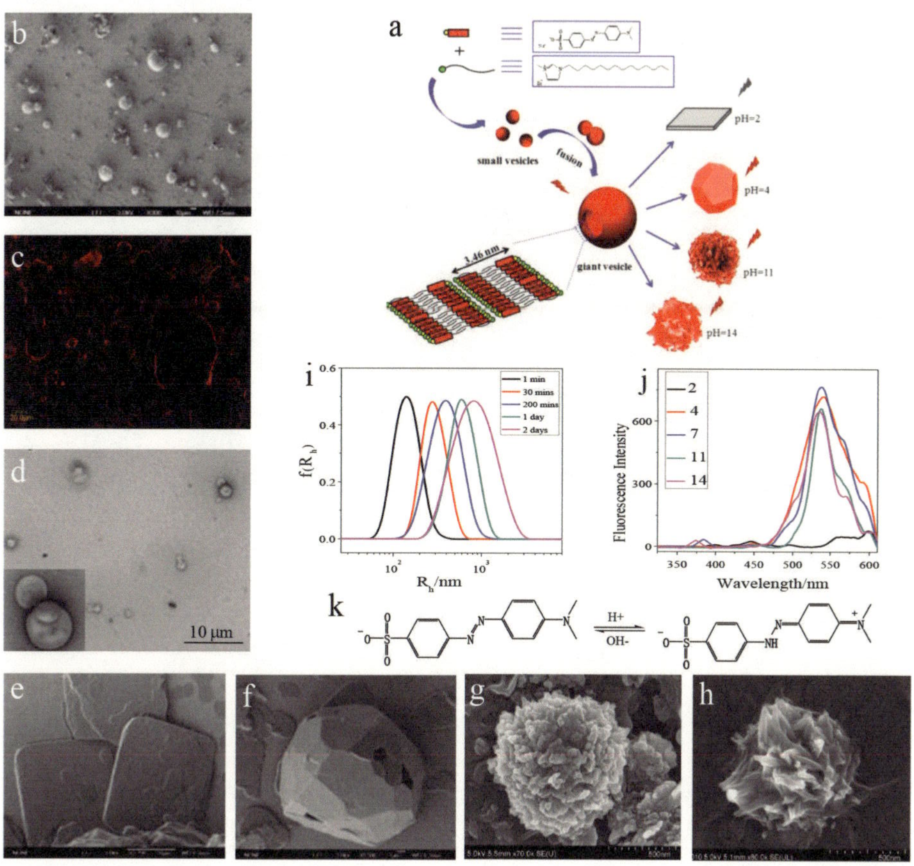

Figure 3. (a) Schematic demonstration of formation of fluorescent giant vesicle and its pH responsivity. (b) SEM and (c) CLSM image of giant vesicles. (d) TEM images of 0.5 mmol L^{-1} MO/C_{14}mimBr after 2 days. SEM images of 0.5 mmol L^{-1} MO/C_{14}mimBr at (e) pH = 2, (f) pH = 4, (g) pH = 11, (h) pH = 14. (i) Time-dependant DLS result of 0.5 mmol L^{-1} MO/C_{14}mimBr. (j) Fluorescence intensity at different pH. (k) pH-dependent mechanism of the MO molecule [26].

state of RhB, minimize aggregation-caused quenching (**Figure 4d**, **e**), enhance the luminescent efficiency of the dye, and improve the luminescence performance, which open up a new way to build soft materials. Besides, the emitting color of system can also be adjusted by changing the molar ratio of NaDC and RhB (**Figure 4f**, **g**) and the, lifetimes of the precipitates increase greatly with an increase in c_{NaDC} (**Figure 4h**). Moreover, they also used another bile salt sodium cholate (SC) and a cationic dye (MB) to obtain bundles of ultra long nanobelts through ISA approach, and the shape and length of the bundles of SC/MB nanobelts could be easily controlled by changing the SC concentration and the aging temperature. Besides, the bundles of ultra long SC/MB nanobelts exhibited efficient electrocatalytic activity toward ascorbic acid (AA) oxidation in phosphate buffer solution (pH = 7.0). This work provides an alternative way to design and fabricate the ultra long belt-like structures with tunable sizes which may also open up a way for the design and development of optical and electronic devices in the potential bioapplications.

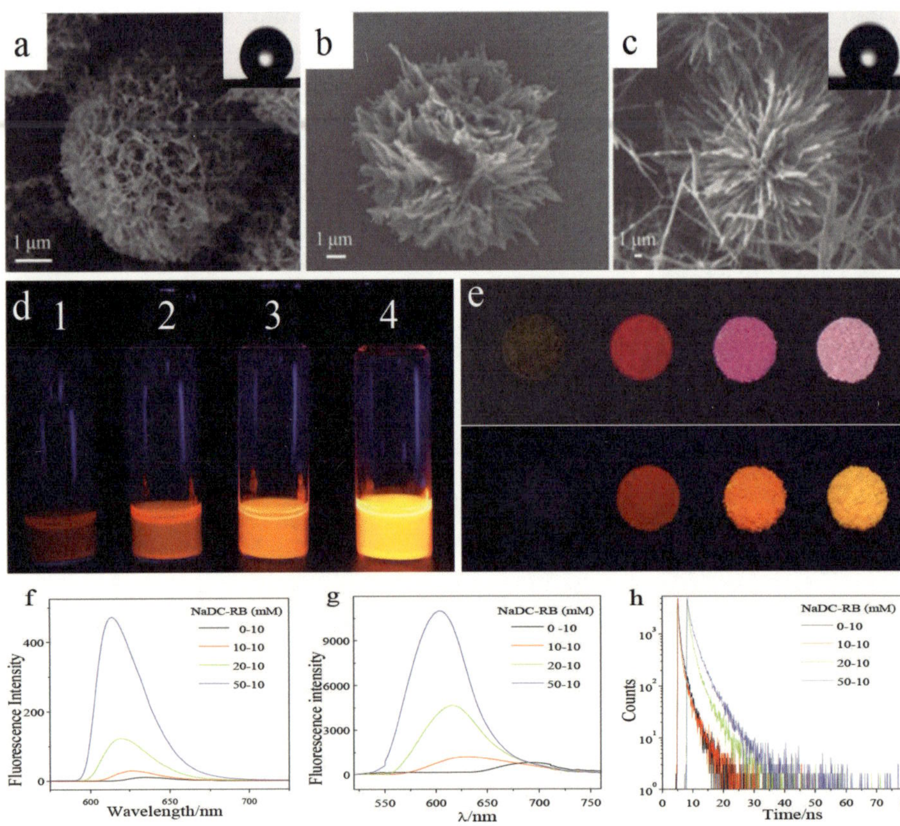

Figure 4. The morphologies of the precipitates of the NaDC/RhB systems when the concentration of RhB was fixed at 10 mmol L^{-1}, while the concentration of NaDC was changed from (a) 10 mmol L^{-1}, (b) 20 mmol L^{-1}, (c) 50 mmol L^{-1}. The inset images of (a) and (c) are the contact angle on the films formed by the sample, respectively. Images of the fluorescence changes in (d): the upper solutions for samples of (1) 10 mmol L^{-1} RhB, (2) 10 mmol L^{-1}NaDC/10 mmol L^{-1} RhB, (3) 20 mmol L^{-1}NaDC/10 mmol L^{-1} RhB and (4) 50 mmol L^{-1}NaDC/10 mmol L^{-1} RhB (excitation at 365 nm), and (e) the powders of the precipitates (upper: under visible light; lower: under UV light, excitation at 365 nm). Changes in the fluorescence spectra of (f) the upper solution and (g) the precipitates obtained for the NaDC/RhB complex samples. (h) Time-resolved fluorescence of the above precipitate samples [33].

Different from conventional planar dyes that always undergo π–π stacking which results innotorious aggregation-caused quenching (ACQ), a new class of dyes with propeller-shaped structure displays aggregation-induced emission (AIE) [34]. However, the drawback of the propeller-shaped topology is that it disfavors self-assembly, so that fluorescent nanostructures based on AIE molecules are still very rarely up to date [35]. In Huang's group, propeller-shaped dye TPE-BPA and surfactant CTAB were chosen to build AIE system [36]. The fluorescence performs the maximum emission at the molar ratio at 1:8 confirming that the optimal interaction between TPE-BPA and CTAB is 1:8 (**Figure 5a**). And in Cryo-TEM observation, it can be found the formation of vesicles with an average size of 145 nm (**Figure 5b**). The hydrophilic heads of TPE-BPA molecules are capable of coordinating with metal ions and with the addition of Zn^{2+} to the TPE-BPA@8CTAB system. It can be found that the fluorescence intensity decreases with the addition of Zn^{2+} and research the plateau at a Zn^{2+}/TPE-BPA ratio of 2 (**Figure 5d, e**), implying that every two coordinating heads share one Zn^{2+} to satisfy the space of an octahedral field. Cryo-TEM reveals the retain of the vesicular structure with smaller size, which was formed by the fission of lager vesicles (**Figure 5c**). The membranes of cancer cells are highly charged compared to the normal ones, the present results indicate that the high electrical charge may accelerate the cell fission and generate a looser molecular packing and increases the releasing rates of hydrophilic drug DOX (**Figure 5f**). The report of this research may help to reveal the mystery behind the easy metastasis of the cancer cells and inspire a novel strategy for cancer therapy.

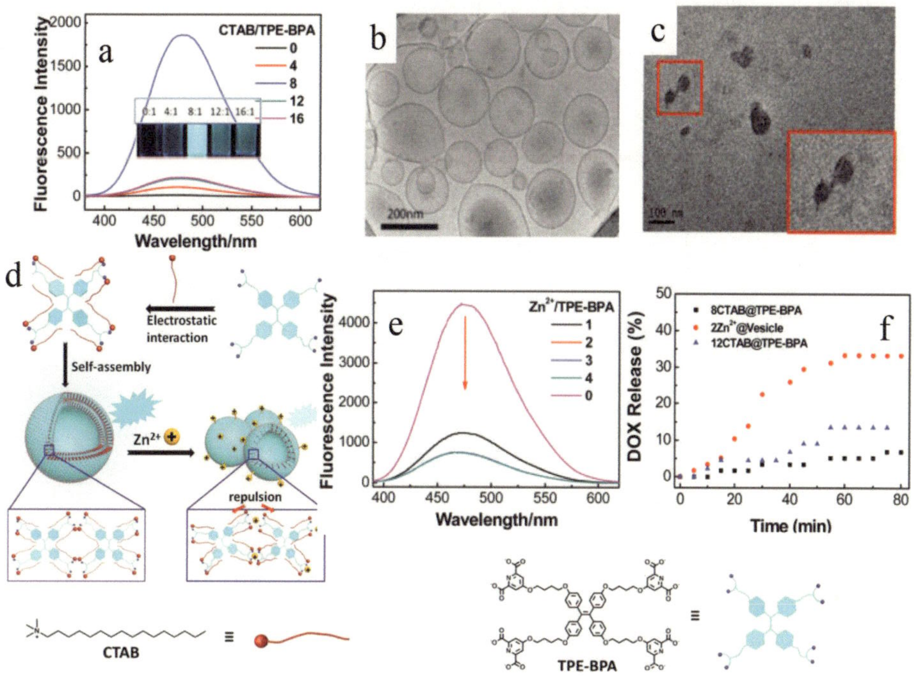

Figure 5. (a) Cryo-TEM image of TPE-BPA@8CTAB vesicle. (b) Fluorescence intensity at different molar ratio of TPE-BPA/CTAB. (c) TEM images of TPE-BPA@8CTAB vesicle triggered by Zn^{2+}. (d) Illustration of self-assembly of TPE-BPA/CTAB vesicles and electrical charge triggered fission of the TPE-BPA/CTAB vesicle. (e) Fluorescence intensity of system with the addition of Zn^{2+} to the TPE-BPA@8CTAB vesicular system. (f) Comparison of the DOX releasing rate in the native TPEBPA@8CTAB vesicle and the charged vesicle of the TPE-BPA@12CTAB, $2Zn^{2+}$@vesicle [35].

3. POM-biomolecule ISA materials

Polyoxometalates (POMs), a large group of metal oxide clusters, represent some of the largest inorganic molecules known so far and have broad applications as catalysts, photoelectronic/magnetic materials, and biologically active materials [37–39]. The POMs are formed by linking metal oxide polyhedra with each other through corner-, edge-, or face-sharing manner, which enable chemists to build POMs with different topologies and sizes. Due to the excess of oxo ligands over metal ions, POMs are usually highly negatively charged, which could interact with positively charged materials by electrostatic interaction. Biomolecule, such as amino acid and polypeptide, is essential for our body [40, 41]. Thus, the self-assembly behavior of POM/biomolecule has potential application in biological field as novel compartments, artificial cell membranes, drug and gene delivery agents [42, 43].

Wu et al. constructed multivalent peptide nanofibers by using short peptides with the synergistic effect of POM clusters [44]. The short peptide L1 adopts a random-coil conformation in aqueous solution, while with the addition of $H_4SiW_{12}O_{40}$ (HSiW), the conformation transition from a random-coil to a β-sheet state (**Figure 6a, e**). In TEM, one-dimensional (1D) nanofibers were observed after the mixing of L1 with HSiW in water (**Figure 6b, c**). The multivalent L1/HSiW nanofibers exhibited significantly enhanced antibacterial efficacy, while the inhibitory ability of L1 or HSiW alone was poor (**Figure 6d**). The ensembly nature of POMs enables the enhancement of the antimicrobial efficacy and biological stability of short peptides in situ.

Xin et al. obtained inorganic-organic hybrid vesicles using POM$Na_9[EuW_{10}O_{36}]\cdot 32H_2O$ (EuW_{10}) and different amino acids (arginine, lysine, histidine, glutamic acid, aspartic acid, leucine, alanine, and phenylalanine) (**Figure 7a, b, g**) [45]. The electrostatic interaction between amino

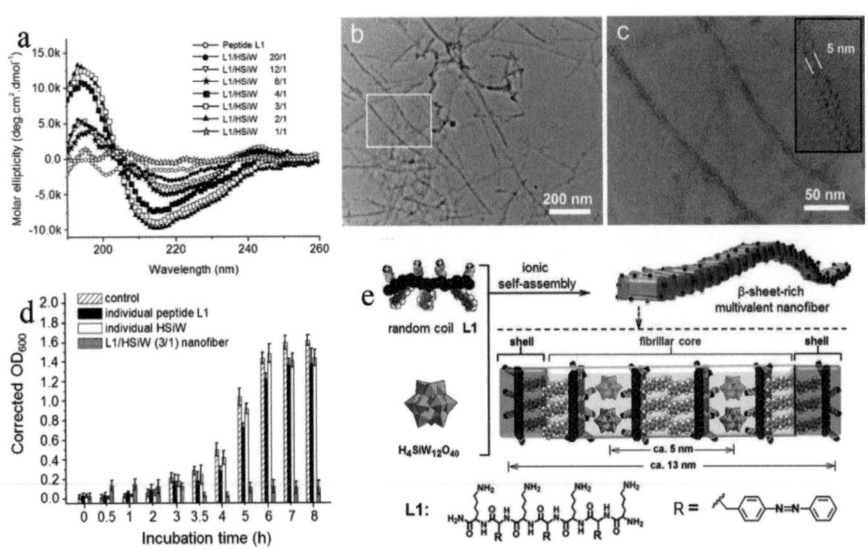

Figure 6. (a) CD spectra of peptide L1 and the L1/HSiW composites at different molar ratios (b, c) TEM image of L1/HSiW at the molar ratio of 3:1. (d) Optical density of *E. coli* with incubation time in the presence of the individual peptide L1, individual HSiW, and L1/HSiW nanofibers. (e) Illustration of self-assembly ofL1/HSiW nanofibers [44].

acid and EuW_{10} plays an important role for the formation of vesicle. While the luminescent property is clear distinction among different amino acids that alkaline amino acids (Arg, Lys and His) enhanced the luminescence, acidic amino acids (Glu, Asp) quenched the luminescence and nonpolar amino acids (Leu, Ala, and Phe) have no obviously influence towards luminescence (**Figure 7c, d**). This is attributed to the strength differences of the electrostatic interactions between amino acids and EuW_{10} cluster. Meaningfully, the Arg/EuW_{10} fluorescent vesicles can be used to detect Dopamine selectively with the detection limit of 3.2 μM on the basis of competition mechanism that dopamine could substitute Arg to form an assembly with EuW_{10} through hydrogen bond interaction between the ammonium group of DA and the oxygen atom of EuW_{10}, leading to fluorescence quenching (**Figure 7e, f**).

Three-dimensional (3D) hierarchical nanostructures have generated large amounts of interests due to their unique peculiar properties and wide range of potential applications. Among them, nanoflower is a fantastic name of some of the nanomaterials which in microscopic images resemble the flowers. Due to the large surface-to-volume ratio compared with that of bulk materials, nanoflowers have many applications in catalysis, magnetism, nanodevices, sensing

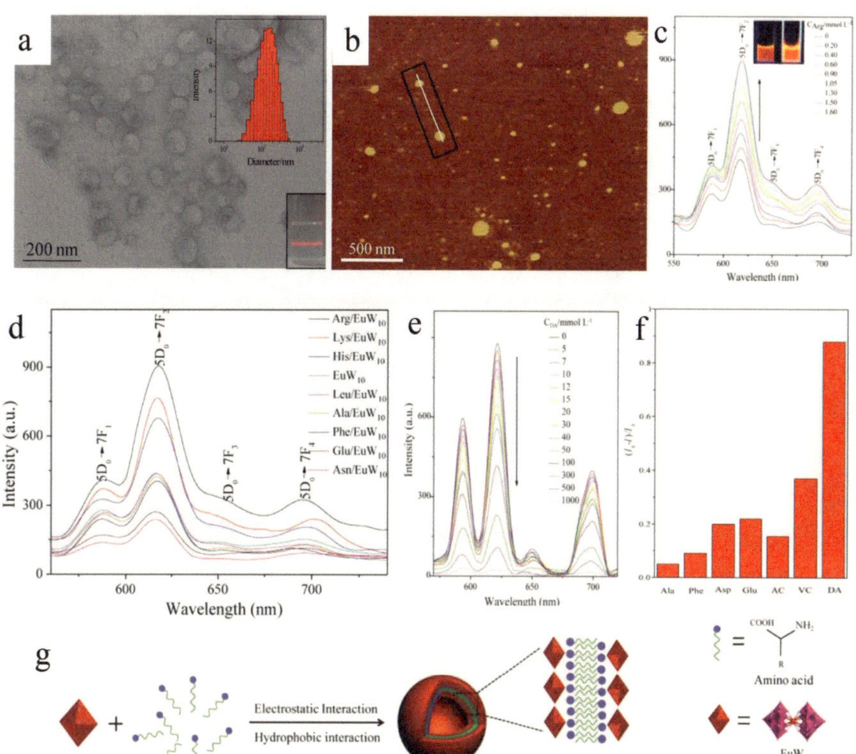

Figure 7. TEM (a) and AFM (b) image of 0.7 mmol L^{-1} EuW_{10}/1.5 mmol L^{-1}Arg (insets are the pictures showing the Tyndall effect under the irradiation by a light beam and the hydrodynamic radius of vesicles observed by DLS). (c) Variation of the fluorescence spectra of EuW_{10} (0.7 mmol L^{-1}) upon the titration of Arg. (d) Fluorescence spectra of EuW_{10} incorporated different amino acids. (e) Variation of the fluorescence spectra of 0.7 mmol L^{-1} EuW_{10}/1.5 mmol L^{-1}Arg with the addition of DA (0–1000 mmol L^{-1}). (f) Fluorescence response of the EuW_{10}/Arg system to biological molecules all the concentrations of the biological molecules were 500 mmol L^{-1}. (g) Schematic illustration of the vesicles formed by EuW_{10} and amino acids [45].

and biosensing, and medicine [46–51]. In Xin's group, inorganic-organic hybrid hierarchical nanoflowers structures were prepared by EuW_{10}/DA on the basis of ISA strategy with the cooperation of H-bonding interaction (**Figure 8a–c**) [52]. The structure of hybrid can be controlled easily by adjusting the ratio of EuW_{10} and DA, and the formation of EuW_{10}/DA hierarchical nanoflowers was monitored by SEM: The protonated DA interact with EuW_{10} by H-bonding and electrostatic interaction, and then, the cores of the flower-like nanostructures were initially formed (**Figure 8d**). Subsequently, more building blocks were stacked into the surface of the cores forming a rough surface and the building blocks aggregated to form the resulted nanoflower (**Figure 8e**). Finally, the hierarchical nanoflowers grew more compact, and the surfaces of the nanopetals became very smooth due to Ostwald ripening in the further growth stage (**Figure 8f**). POMs have been used as catalysts for the oxidation of a variety of compounds such as alkenes, alcohols, sulfides and dyes [53]. In this work, the author calcinated the EuW_{10}/DA nanoflower at 350°C for 2 h to get a pure EuW_{10} framework, and the porous EuW_{10} performs excellent degradation ability for MO with the exist of H_2O_2, the catalyst exhibits only a little loss of photocatalytic activity after six recycles for the degradation of MO, implying that the calcinated porous EuW_{10} can be used as an excellent photocatalyst for multiple cycles to catalyze MO and have potential application for sewage treatment (**Figure 8g–i**). The schematic illustration of the assembly process was shown in **Figure 9**.

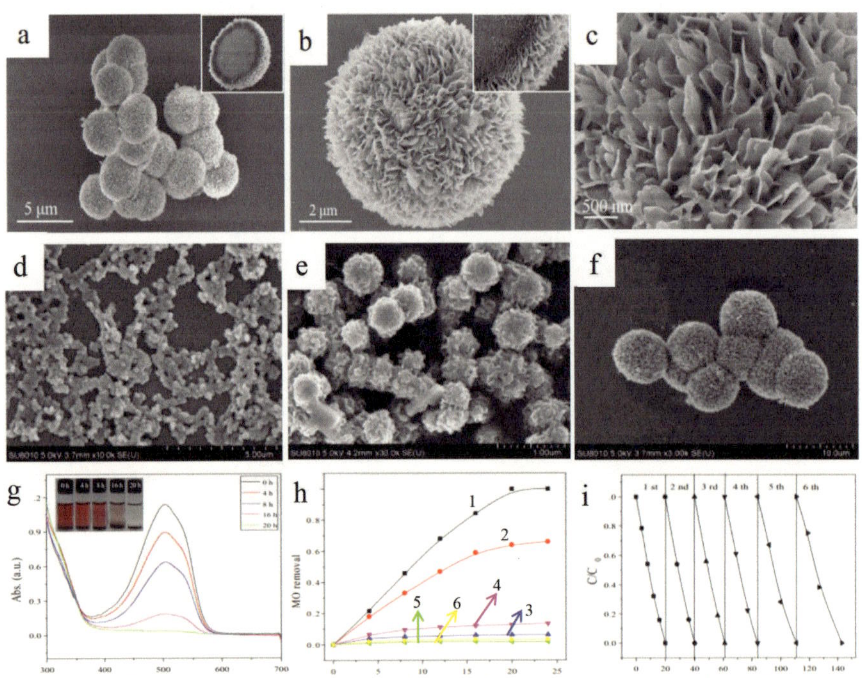

Figure 8. (a, b, c) SEM images of nanoflower formed by 2 mg mL^{-1} EuW_{10}/2 mg mL^{-1} DA at different scales. The formation process of EuW_{10}-DA nanoflowers with different incubation time (d) 1 min, (e) 0.5 h, and (f) 2 h. (g) The UV-vis curves of the degradation of MO by calcinated nanoflower. (h) MO degradation over time with different substances: (h-1) calcinated porous EuW_{10}/H_2O_2, (h-2) EuW_{10}-DA hybrid nanoflowers/H_2O_2,(h-3) EuW_{10}/H_2O_2, (h-4) DA/H_2O_2, (h-5) H_2O_2 and (h-6) calcinated porous EuW_{10}. (i) The recycling experiment for the degradation of MO using calcinated porous EuW_{10}/H_2O_2 system [52].

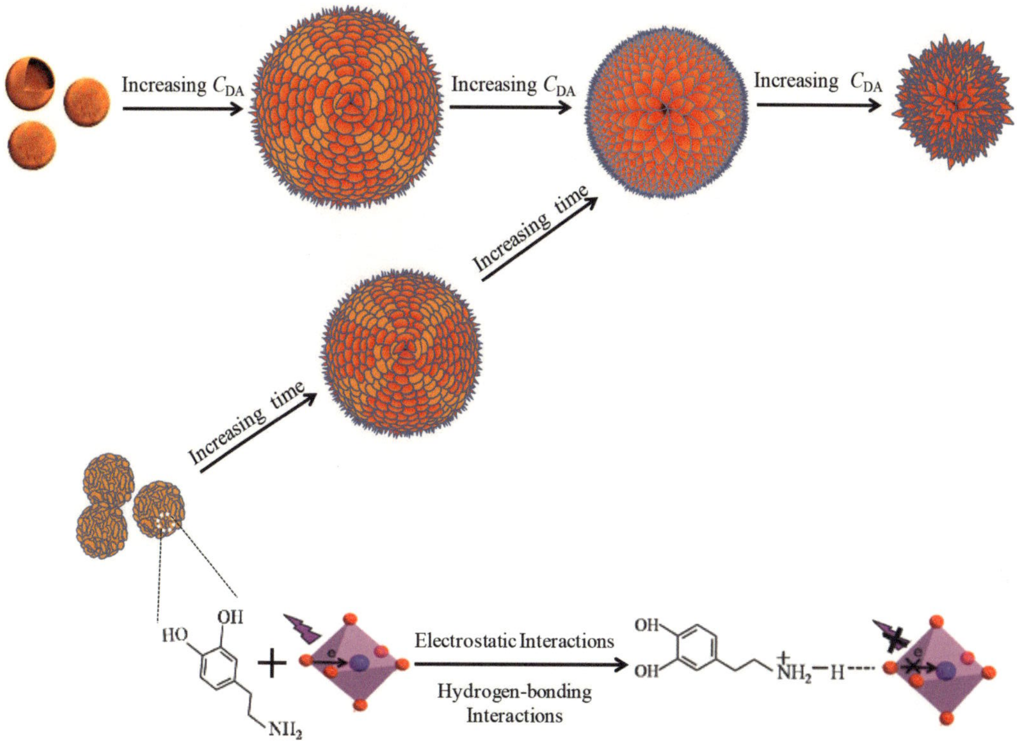

Figure 9. The schematic illustration of the assembly process for EuW_{10}-DA hybrid nanoflowers [52].

4. Conclusion and outlook

In a conclusion, ISA strategy on the basis of electrostatic interaction is a facile and convenient method to prepared complex and hierarchical materials for a wide range of applications without tangle some covalent functionlization. Although many studies have focused on the design and preparation of advanced materials using ISA strategy in the past few years, the successful translation of these laboratory innovations to tackle specific problems in the real world remains a great challenge. It is anticipated that preparing functional materials through ISA strategy is a promising pathway to design and generate new nanoassemblies with unique properties and will certainly play a significant role in the future of material science.

Acknowledgements

We gratefully acknowledge the financial support from the National Natural Science Foundation of China (21573130,21173128) and the Young Scholars Program of Shandong University (2016WLJH20).

Author details

Jinglin Shen[1], Shiling Yuan[1] and Xia Xin[1,2*]

*Address all correspondence to: xinx@sdu.edu.cn

1 Key Laboratory for Colloid and Interface Chemistry of Education Ministry, School of Chemistry and Chemical Engineering, Shandong University, Jinan, P. R. China

2 National Engineering Technology Research Center for Colloidal Materials, Shandong University, Jinan, P. R. China

References

[1] Gan Q, Ferrand Y, Bao C, Kauffmann B, Grelard A, Jiang H, Huc I. Helix-rod host-guest complexes with shuttling rates much faster than disassembly. Science. 2011; 331: 1172–1175. doi: 10.1126/science.1200143

[2] Lee C C, Grenier C, Meijer E W, Schenning A P H J. Preparation and characterization of helical self-assembled nanofibers. Chem. Soc. Rev. 2009; 38: 671–683. doi: 10.1039/B800407M

[3] Vera F, Serrano J L, Sierra T. Twists in mesomorphic columnar supramolecular assemblies. Chem. Soc. Rev. 2009; 38: 781–796. doi: 10.1039/B800408K

[4] Huang Z, Kang S K, Banno M, Yamaguchi T, Lee D, Seok C, Yashima E, Lee M, Huang Z, Kang S K, Banno M, Yamaguchi T, Lee D, Seok C, Yashima E, Lee M. Pulsating tubules from noncovalentmacrocycles. Science. 2012; 337: 1521–1526. doi: 10.1126/science.1224741

[5] Zhang W, Jin W, Fukushima T, Saeki A, Seki S, Aida T. Supramolecular linear heterojunction composed of graphite-like semiconducting nanotubular segments. Science. 2011; 334: 340–343. doi: 10.1126/science.1210369

[6] Cordier P, Tournilhac F, Soulie-Ziakovic C, Leibler L. Self-healing and thermoreversible rubber from supramolecular assembly. Nature. 2008; 451: 977–980. doi: 10.1038/nature06669

[7] Aida T, Meijer E W, Stupp S I. Functional supramolecular polymers. Science. 2012; 335: 813–817. doi: 10.1126/science.1205962

[8] Zeng F, Zimmerman S C. Dendrimers in supramolecular chemistry: from molecular recognition to self-assembly. Chem. Rev. 1997; 97: 1681–1712. doi: 10.1021/cr9603892

[9] Chakrabarty R, Mukherjee P S, Stang P J. Supramolecular coordination: self-assembly of finite two- and three-dimensional ensembles. Chem. Rev. 2011; 111: 6810–6918. doi: 10.1021/cr200077m

[10] Yu G, Jie K, Huang F. Supramolecular amphiphiles based on host–guest molecular recognition motifs. Chem. Rev. 2015; 115: 7240–7303. doi: 10.1021/cr5005315

[11] Tan L, Liu Y, Ha W, Ding S, Peng S, Zhang S, Li J. Stimuli-induced gel–sol transition of multi-sensitive supramolecular β-cyclodextrin grafted alginate/ferrocene modified pluronic hydrogel. Soft Matter, 2012; 8: 5746–5749. doi: 10.1039/C2SM25084E

[12] Zhu Y, Liu L, Du J. Probing into homopolymer self-assembly: how does hydrogen bonding influence morphology? Macromolecules, 2013; 46: 194–203. doi: 10.1021/ma302176a

[13] Tahara K, Lei S B, Adisoejoso J, Feyter S De, Tobe Y. Supramolecular surface-confined architectures created by self-assembly of triangular phenylene–ethynylenemacrocycles via van der Waals interaction. Chem. Commun. 2010; 46: 8507–8525. doi: 10.1039/C0CC02780D

[14] Zhu Y, Fan L, Yang B, Du J. Multifunctional homopolymer vesicles for facile immobilization of gold nanoparticles and effective water remediation. ACS Nano. 2014; 8: 5022–5031. doi: 10.1021/nn5010974

[15] Li Q, Chen X, Wang X, Zhao Y. Ionic self-assembled wormlike nanowires and their cyclodextrin inclusion-tuned transition. J. Phys. Chem. B, 2010; 114: 10384–10390. doi: 10.1021/jp104801m

[16] Liu T, Tian W, Zhu Y. How does a tiny terminal alkynyl end group drive fully hydrophilic homopolymers to self-assemble into multicompartment vesicles and flower-like complex particles? Polym. Chem. 2014; 5: 5077–5088. doi: 10.1039/C4PY00501E

[17] Faul C F J, Antonietti M. Ionic self-assembly: facile synthesis of supramolecular materials. Adv. Mater. 2003; 15: 673–683. doi: 10.1002/adma.200300379

[18] Thool G S, Narayanaswamy K, Venkateswararao A, Naqvi S, Gupta V, Chand S, Vivekananthan V, Koner R R, Krishnan V, Singh S P. Highly directional 1D-supramolecular assembly of new diketopyrrolopyrrole based gel for organic solar cell applications. Langmuir. 2016; 32: 4346–4351. doi: 10.1021/acs.langmuir.6b00846

[19] Faul C F J. Ionic self-assembly for functional hierarchical nanostructured materials. Acc. Chem. Res. 2014; 47: 3428–3438. doi: 10.1021/ar500162a

[20] Adak S, Datta S, Bhattacharya S, Banerjee R. Imidazolium based ionic liquid type surfactant improves activity and thermal stability of lipase of Rhizopusoryzae. J. Mol. Catal. B: Enzym. 2015; 119: 12–17. doi: 10.1016/j.molcatb.2015.05.010

[21] Wang B, Song A, Feng L, Ruan H, Li H, Dong S, Hao J. Tunable amphiphilicity and multifunctional applications of ionicliquid-modified carbon quantum dots. ACS Appl. Mater. Interfaces 2015; 7: 6919–6925. doi: 10.1021/acsami.5b00758

[22] Guan Y, Zakrevskyy Y, Stumpe J, Antonietti M, Faul C F. Perylenediimide-surfactant complexes: thermotropic liquid-crystalline materials via ionic self-assembly. Chem. Commun. 2003; 7: 894–895. doi: 10.1039/B211753C

[23] Zhao M, Zhao Y, Zheng L, Dai C. Construction of supramolecular self-assembled microfibers with fluorescent properties through a modified ionic self-assembly (ISA) strategy. Chem. Eur. J. 2013; 19: 1076–1081. doi: 10.1002/chem.201203062

[24] Faul C F J, Antonietti M. Facile synthesis of optically functional, highly organized nanostructures: dye–surfactant complexes. Chem. Eur. J. 2002; 8: 2764–2768. doi: 10.1002/1521-3765(20020617)8:12<2764::AID-CHEM2764>3.0.CO;2-X

[25] Shen J, Xin X, Liu T. Ionic self-assembly of a giant vesicle as a smart microcarrier and microreactor. Langmuir. 2016; 32: 9548–9556. doi: 10.1021/acs.langmuir.6b01829

[26] Shen J, Xin X, Liu G. Fabrication of smart pH-responsive fluorescent solid-like giant vesicles by ionic self-assembly strategy. J. Phys. Chem. C. 2016; 120: 27533–27540. doi: 10.1021/acs.jpcc.6b08140

[27] Fan Y, Zhang D, Wang J, Jin H, Zhou Y, Yan D. Preparation of anion-exchangeable polymer vesicles through the selfassembly of hyperbranched polymeric ionic liquids. Chem. Commun. 2015; 51: 7234–7237. doi: 10.1039/C5CC01802A

[28] Galantini L, Gregorio M C, Gubitosi M, Travaglini L, Tato J V, Jover A, Meijide F, Tellini V H S, Pavel N V. Bile salts and derivatives: rigid unconventional amphiphiles as dispersants, carriers and superstructure building blocks. Curr. Opin. Colloid Interface Sci., 2015; 20: 170–182. doi: 10.1016/j.cocis.2015.08.004

[29] Schefer L, Sánchez-Ferrer A, Adamcik J, Mezzenga R. Resolving self-assembly of bile acids at the molecular length scale. Langmuir, 2012; 28: 5999–6005. doi: 10.1021/la300384u

[30] Tung S H, Huang Y E, Raghavan S R. A new reverse wormlike micellar system: mixtures of bile salt and lecithin in organic liquids. J. Am. Chem. Soc. 2006; 128: 5751–5756. doi: 10.1021/ja0583766

[31] Sun X, Xin X, Tang N, Guo L, Wang L, Xu G. Manipulation of the gel behavior of biological surfactant sodium deoxycholate by amino acids. J. Phys. Chem. B. 2014; 118: 824–832. doi: 10.1021/jp409626s

[32] Reschly E J, Ai N, Ekins S, Welsh W J, Hagey L R, Hofmann A F, Krasowski M D. Evolution of the bile salt nuclear receptor FXR in vertebrates, J. Lipid Res. 2008; 49: 1577–1587. doi: 10.1194/jlr.M800138-JLR200

[33] Song Z, Xin X, Shen J. Reversible controlled morphologies switching between porous microspheres and urchin-like microcrystals for NaDC/RhB self-assembly and their multifunctional applications. J. Mater. Chem. C. 2016; 4: 8439–8447. doi: 10.1039/C6TC02329K

[34] Hong Y, Lam J W Y, Tang B Z. Aggregation-induced emission. Chem. Soc. Rev. 2011; 40: 5361–5388. doi: 10.1039/C1CS15113D

[35] Li J, Liu K, Han Y, Tang B, Huang J, Yan Y. Fabrication of propeller-shaped supra-amphiphile for construction of enzyme-responsive fluorescent vesicles. ACS Appl. Mater. Interfaces. 2016; 8: 27987–27995. doi: 10.1021/acsami.6b08620

[36] Li J, Shi K, Drechsler M, Tang B, Huang J, Yan Y. A supramolecular fluorescent vesicle based on a coordinating aggregation induced emission amphiphile: insight into the role of electrical charge in cancer cell division. Chem. Commun. 2016; 52: 12466–12469. doi: 10.1039/C6CC06432A

[37] Hill C L. Introduction: polyoxometalates multicomponent molecular vehicles to probe fundamental issues and practical problems. Chem. Rev. 1998; 98: 1–2. doi: 10.1021/cr960395y

[38] Long D L, Burkholder E, Cronin L. Polyoxometalate clusters, nanostructures and materials: from self assembly to designer materials and devices. Chem. Soc. Rev. 2007; 36: 105–121. doi: 10.1039/B502666K

[39] Yin P, Li D, Liu T. Solution behaviors and self-assembly of polyoxometalates as models of macroions and amphiphilic polyoxometalate–organic hybrids as novel surfactants. Chem. Soc. Rev. 2012; 41: 7368–7383. doi: 10.1039/C2CS35176E

[40] Shimizu T, Masuda M, Minamikawa H. Supramolecular nanotube architectures based on amphiphilic molecules. Chem. Rev. 2005; 105: 1401–1443. doi: 10.1021/cr030072j

[41] Landsmann S, Luka M, Polarz S. Bolaform surfactants with polyoxometalate head groups and their assembly into ultra-small monolayer membrane vesicles. Nat. Commun. 2012; 3: 1299. doi: 10.1038/ncomms2321

[42] Stanish I, Lowy D A, Hung C W, Singh A. Vesicle-based rechargeable batteries. Adv. Mater. 2005; 17: 1194–98. doi: 10.1002/adma.200401132

[43] Duan Q, Cao Y, Li Y, Hu X, Xiao T, Lin C. pH responsive supramolecular vesicles based on water-soluble pillar[6]-arene and ferrocene derivative for drug delivery. J. Am. Chem. Soc. 2013; 135: 10542–10549. doi: 10.1021/ja405014r

[44] Li J, Chen Z, Zhou M. Polyoxometalate-driven self-assembly of short peptides into multivalent nanofibers with enhanced antibacterial activity. Angew. Chem. Int. Ed. 2016; 128: 2638–2641. doi: 10.1002/ange.201511276

[45] Zhang H, Guo L, Xie Z. Tunable aggregation-induced emission of polyoxometalates via amino acid-directed self-assembly and their application in detecting dopamine. Langmuir. 2016; 32: 13736–13745. doi: 10.1021/acs.langmuir.6b03709

[46] ZengM, LiY, LiuF, YangY, MaoM, ZhaoX. Cu doped OL-1 nanoflower: a UV-vis-infrared light-driven catalyst for gas-phase environmental purification with very high efficiency. Appl. Catal. B: Environ. 2017; 200: 521–529. doi: 10.1016/j.apcatb.2016.07.042

[47] Huang K, Liu Y, Wang L. Molybdenum disulfide nanoflower-chitosan-Au nanoparticles composites based electrochemical sensing platform for bisphenol A determination. J. Hazard. Mater. 2014; 276: 207–215. doi: 10.1016/j.jhazmat.2014.05.037

[48] He S, Hu C, Hou H, Chen W. Ultrathin MnO_2 nanosheets supported on cellulose based carbon papers for high-power supercapacitors. J. Power. Sources. 2014; 246: 754–761. doi: 10.1016/j.jpowsour.2013.08.038

[49] Wang D, Pan Z, Wu Z, Wang Z, Liu Z. Hydrothermal synthesis of MoS_2 nanoflowers as highly efficient hydrogen evolution reaction catalysts. J. Power Sources. 2014; 264: 229–234. doi: 10.1016/j.jpowsour.2014.04.066

[50] Liu Y, Jiao Y, Zhang Z, Qu F, Umar A, Wu X. Hierarchical SnO_2 nanostructures made of intermingled ultrathin nanosheets for environmental remediation, smart gas sensor, and supercapacitor applications. ACS Appl. Mater. Interfaces. 2014; 6: 2174–2184. doi: 10.1021/am405301v

[51] Hu L, Ren Y, Yang H, Xu Q. Fabrication of 3D hierarchical MoS_2/polyaniline and MoS_2/C architectures for lithium-ion battery applications. ACS Appl. Mater. Interfaces. 2014; 6: 14644–14652. doi: 10.1021/am503995s

[52] Zhang H, Guo L Y, Jiao J. Ionic Self-assembly of Polyoxometalate-Dopamine Hybrid Nanoflowers with Excellent Catalytic Activity for Dyes. ACS Sustainable Chem. Eng. 2017; 5: 1358–1367. DOI: 10.1021/acssuschemeng.6b01805

[53] Mizuno N, Yamaguchi K, Kamata K. Epoxidation of olefins with hydrogen peroxide catalyzed by polyoxometalates. Coordin. Chem. Rev. 2005; 249: 1944–1956. DOI:10.1016/j.ccr.2004.11.019

Chapter 4

Chiral Solvation Induced Supramolecular Chiral Assembly of Achiral Polymers

Wei Zhang, Yin Zhao and Lu Yin

Additional information is available at the end of the chapter

http://dx.doi.org/10.5772/67700

Abstract

To date, liquid crystal chirality, mechanophysical chirality, circularly polarized photon chirality, gelation and chiral solvation are all feasible candidates to generate optically active polymers and supramolecular chirality when employing achiral molecules as starting substances. Among this, chiral-solvation-induced chirality is one of the dominant methods for construction of chirality from achiral sources, such as achiral poly(*n*-hexyl isocyanate) (PHIC), π-conjugated polymers, oligo(*p*-phenylenevinylene), polyacetylenes, σ-conjugated polysilanes and side-chain polymers. Supramolecular chirality is well established through their intra- or inter-molecular noncovalent interactions, such as van der Waals, CH/π, dipole-dipole interactions, hydrogen bonding and metal-ligand coordinating interactions. Compared with the traditional methods, this strategy avoids the use of expensive chiral reagents and also expands the scope towards challenging substrates. This chapter highlights a series of studies that include: (i) the development-historical background of chiral solvent induction strategy; (ii) the chiral-solvation-induced chirality in small molecules and oligomers; and (iii) recent developments in polymers, especially in π-conjugated polymers and σ-conjugated polymers.

Keywords: optical activity, supramolecular chirality, chiral solvation, self-assembly, circular dichroism, circularly polarized luminescence

1. Introduction

As early as the second half of the nineteenth century, many scientists have long thought that the intrinsic biomolecular homochirality found in the living world is the origin of life on earth, since inherent optical activity exists inside all living organisms [1–14]. For example, typical fundamental components of our body, DNA and polypeptide biopolymers, consist of *D*-ribose and *L*-amino acid building blocks with the same handedness, respectively. The absolute enantiopurity inside living organisms is of prime importance for many biological processes involving molecular recognition and replication, enzymatic catalysis, heritable characters, and pharmaceutical and toxicological activities [15]. Therefore, the studies on chirality were actually meant to explore life itself, its origin and nature.

Lord Kelvin first coined the term chirality as 'I call any geometrical figure, or group of points, chiral, and say it has chirality if its image in a plane mirror, ideally realized, cannot be brought to coincide with itself'. Many pioneering concepts, such as 'optical activity', 'asymmetry', 'enantiomers' and 'chirality', also have been proposed since Pasteur's discovery of handedness in molecules. Research on chirality increased the demand of chiral materials and motivated more studies on synthetic chiral materials, especially chiral polymers, which have great promising applications in asymmetric catalysis, chiral resolution, chirality sensor and optoelectronic materials [16–24]. In general, most molecules or polymers possessing handed stereogenic centres and/or handed stereogenic bonds are considered to be optically active or chiral. This kind of chirality is attributed to straightforward chiral induction biases. Nowadays, traditional methods for preparing these chiral polymers have encountered many problems like the use of expensive chiral monomers or complex asymmetric polymerization processes. It is worth noting that, even if many substances have no optical activity, it does not mean that they cannot be chiral. They may exist as racemic mixtures or exist as time-averaged structures in a mirror-symmetric potential energy surface. External influence of molecular species, including stereogenic centres and/or stereogenic bonds, may indirectly induce their chiroptical activity. What's more, optical activity should be recognized as observable and measurable chiroptical signals in the ground and photoexcited states, but not as the typical defined chirality with chiral structure itself. Introducing asymmetry into optically inactive sources to obtain supramolecular chirality has gained increasing interest due to the appeal of avoiding of tedious synthesis of chiral polymers, as well as the possibility of several chiroptical applications including memory and switching.

Nowadays, methods for introducing asymmetry have been expanded widely, for example, using asymmetric liquid crystal field, supramolecular interactions with small chiral molecules, chiral circular polarized light and chiral solvation [21, 24–28]. Among these, chiral solvation method by simply using chiral solvent molecules as chiral source provides a relatively greener way to prepare optically active polymers. In this way, achiral macromolecules are surrounded by numerous chiral solvent molecules, and noncovalent supramolecular interactions (acid-base interaction, hydrogen bond, metal-ligand interaction and van der Walls force) existing between them will probably produce optical activity. Although these chiral induction biases are weak, chirality transfer from small molecules to macromolecules could be successfully realized through this way. Also, chiral induction biases can be amplified with the help of solvent quantity, subsequently obtaining supramolecular chirality. The concept of supramolecular chemistry was first delineated by Lehn as "the chemistry of molecular assemblies and of intermolecular bond" [29]. Supramolecular chemistry is closely allied to self-assembly, which has been defined as the spontaneous organization of pre-existing disordered components into ordered structures or pattern, as a consequence of specific interactions among those components themselves without external direction. Both supramolecular chemistry and molecular self-assembly are related to noncovalent interactions. They are great

http://dx.doi.org/10.5772/67700

essential parts of biological systems, and exist everywhere such as the transfer and storage of genetic information in nucleic acids, and the self-folding of proteins into more functionally active molecular machines. These biological molecular self-assembly processes produce supramolecular chirality, due to the special spatial arrangements of the biological molecules. In a self-assembled system, both chiral and achiral components can be used to produce supramolecular chirality.

To detect this kind of supramolecular chirality in the ground states, circular dichroism (CD) and optical rotation dispersion (ORD) are used to investigate the electronic transition properties of the chiral species [30]. Vibrational circular dichroism (VCD) and Raman optical activity are used to investigate their vibronic transition properties [31, 32]. On the contrast, circularly polarized luminescence (CPL) provides us their structure information of chiral species in the photoexcited states [33]. These molecular structures are also predictable computationally by using the time-dependent density functional theory (DFT). The various processes of chiral supramolecular assembly can be monitored by UV-visible spectrophotometry, circular dichroism (CD) and fluorescent spectroscopy. Furthermore, their micro morphologies (nanoparticles or nanofibers) can be investigated by Atomic Force Microscopy (AFM), Transmission Electron Microscope (TEM) and Scanning Electron Microscope (SEM) technologies. This chapter highlights a series of studies that include: (i) the development-historical background of chiral solvent induction strategy; (ii) the chiral-solvation-induced chirality in small molecules and oligomers; and (iii) recent developments in polymers, especially in π-conjugated polymers and σ-conjugated polymers.

2. Chirality of small molecules and oligomers induced by chiral solvation

2.1. Induced circular dichroism for small molecules

It is well known that the induced optical activity for the absorption band(s) of the achiral species is defined as "induced circular dichroism (ICD)" [34]. One of the earliest observations was made when small achiral or CD-silent molecules were dissolved in chiral solvents, and the mirror symmetry of the chromophoric substrate was successfully broken by the solvation. In 1898, Kipping and Pope reported the first stirred chiral crystallization of aqueous $NaClO_3$ solutions in *D*-dextrose, *D*-mannitol and *D*-dulcitol [35–37]. Although the aqueous solution of sodium chlorate is not optically active, the crystals obtained showed strong bias towards formation of one enantiomorphic form due to the successful chirality transfer from sugar solute. In isotropic solutions, the chiral solvation method applied in several small CD-silent molecules was then investigated [38]. As early as 1965, Mason et al. observed for the first time ICD bands for the *d-d* transitions of $[Co(NH_3)_6](ClO_4)_3$ in aqueous diethyl-(+)-tartrate solutions, according to the outer-sphere coordination between them [39]. What's more, Bosnich et al. found ICD effects not only for the *d-d* transitions of $[PtC1_4]^{2-}$ but also for the n-π^* transitions of benzyl and benzophenone in (*S*,*S*)-2,3-butanediol (**Figure 1**) [40]. The ICD effects for these aromatic groups were ascribed to the inherently twisting conformations of benzyl groups due to their H-H repulsion [41]. Further study on ICD effects has been reported by

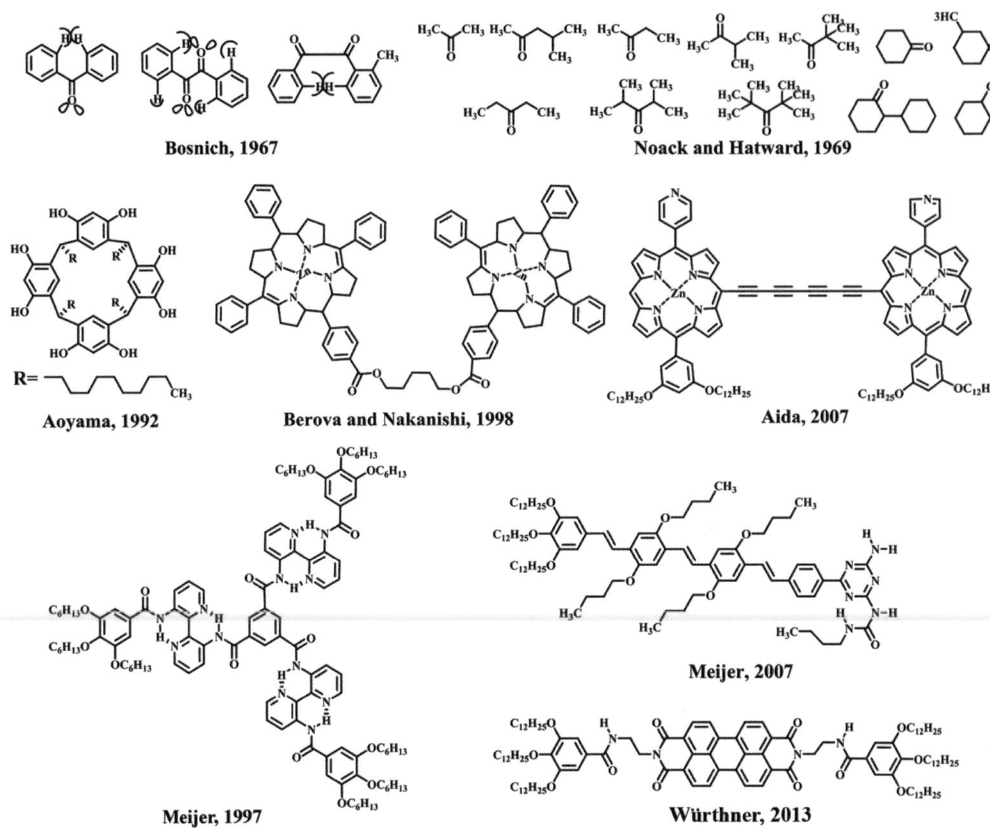

Figure 1. Optically active small molecules and oligomers induced by chiral salvation.

Hayward et al. for the *n*-*π** transitions of 10 symmetric and racemic aliphatic ketones in six chiral tetrahydrofuranols (**Figure 1**) [42]. Later, a molecular complex with 1:1 molar ration between achiral-saturated ketones and chiral *L*-menthol was proposed to exist in the solution by investigating the concentration dependence of ICD bands [43]. This was possibly ascribed to chiral OH/O interactions. Such induced CD (ICD) effects have so far been observed in a number of situations, and the intermolecular interactions such as hydrogen bonding, van der Waals interactions, or ionic coupling between chiral species and achiral ones are contributed to the induction of the optical activity for the electronic transitions of achiral species.

2.2. Induced circular dichroism for oligomers

Another most frequently studied type of ICD effects is between achiral or CD-silent oligomers and chiral solvents, due to their intra- or inter-molecular interactions. In 1992, Aoyama et al. prepared a hydrophobic resorcinol cyclic tetramer (host molecule) by reaction between resorcinol and dodecanal (**Figure 1**) [44, 45]. Through cooperative CH-*π* and hydrogen-bonding effect, this achiral, chromophoric host molecule efficiently formed soluble complexes with chiral, nonchromophoric guest molecules, such as various glycols and sugars, (*R*)-/(*S*)-2-pentanol,

(*R*)-/(*S*)-1-phenylethanol, *D*-/*L*-menthol, epicholestanol and cholesterol. The desymmetrization of the host molecule upon host-guest complexation enabled the coupling of the transition moments generated by the aromatic rings, giving rise to not only ICD effects but also exciton-coupled CD spectra. This kind of observed couplet was also further applied to determine the absolute configuration of the guest molecules. The complexation between CD-silent zinc porphyrins and chiral amines or alcohols is another important intermolecular interaction applied to prepare optically active complexes. Berova et al. have designed an achiral zinc bis-porphyrin linked with a long pentamethylene diester (host molecule) as shown in **Figure 1** [46–48]. By relatively intense coordination interactions between zinc porphyrin and nitrogen or oxygen, the host molecules successfully bound plenty of chiral diamines, amino alcohols and amino acids and showed clear ICD effects, due to the formation of CD-active 1:1 and 1:2 host-guest complexes. Borovkov, Inoue and coworkers have synthesized a CD-silent zinc bis-porphyrin rotamer linked with a shorter 1,2-ethane spacer (**Figure 1**) [49–51]. When it binds chiral secondary amines or chiral secondary alcohols through ligand-to-metal coordination, the rotamer shows efficient switch between *syn*- and *anti*-conformations. This kind of supramolecular chirality is caused by the steric interactions between the 3,7-ethyl groups of the porphyrin and the substituents of the ligand. Tsuda and Aida designed a zinc bis-porphyrin rotamer containing a rigid linker and pyridine substituent [52]. When it is dissolved in asymmetric hydrocarbons, such as (*S*)- and (*R*)-limonene, it is capable of self-assembling into a twisted box-shaped tetramer through multipoint non-covalent interactions and gives rise to the chiral supramolecular assembly. The homochirality of the self-assembled tetramer was then characterized by analysing its CD spectra as functions of limonene enantiopurity and the time-dependent CD change in a dilute condition. Also, the obtained tetrameric assembly is enantiomerically enriched and optically active, confirming the successful preparation of supramolecular chiroptical sensor for chiral limonene. And from the CD spectra of the porphyrin box, the optical purity and absolute configuration of limonene can be determined.

Aside from breaking the inner mirror symmetry, the induced optical activity of achiral oligomers can also be accomplished by forming supramolecular assemblies when the oligomers are aggregating in chiral solvents. Meijer et al. presented pioneering studies on the construction of supramolecular chirality for self-assembled C_3 symmetric disc-shaped molecules and achiral oligo(*p*-phenylenevinylene) derivatives (OPVs) through preferential chiral solvation (**Figure 1**) [53–55]. In their work, the chiral inducers, such as (*S*)- and (*R*)-citronellic acid, (*S*)- and (*R*)-citronellol and (*S*)-2,6-dimethyloctane, showed chiroptical induction ability and promised homochirality of the assemblies through chiral OH/N, OH/O and CH/π interactions. For achiral OPVs equipped with ureidotriazine arrays and *n*-butoxy side chains, chiral solvation gives rise to quadruple H-bonded dimers and subsequently forms well-defined supramolecular stacks by cooperative π-π stacking [54, 55]. Würthner et al. demonstrated that chiral solvent, (*S*)- and (*R*)-limonene, successfully generated one-dimensional helical nanofibres from optically inactive amide-functionalized perylene bisimide derivatives (PBI) by synergic effects of hydrogen-bonding interactions and π-π stacking (**Figure 1**) [56]. The presence of reversed CD effects of this PBI gelator in (*S*)- and (*R*)-limonene reveals that helically packing of the chromophore is successfully directed by the chiral environment, which was also proven by the chiral bias towards homochiral aggregates observed by AFM studies.

3. Chirality of polymers induced by chiral solvation

Apparently, compared with small organic molecules, optically active polymers possess better film-processing ability, thermodynamic stability and particular physical and chemical properties. These polymers are always prepared by polymerization with chiral monomers, asymmetric polymerization with achiral monomers, or chiral self-assembly processes from optically inactive sources. The latest method has been studied intensively and been recognized as one of the most promising ways for preparing optically active polymers. Nowadays, inter- or intra-molecular interactions (like van der Waals, CH/π, hydrogen bonding, coulombic, charge-transfer and metal-ligand coordinating interactions) have been used to facilitate the chirality induction of optically active polymers from optically inactive polymers. As early as in 1993, Green et al. observed for the first time a macromolecular conformational change driven by a minute chiral solvation energy [26, 57]. The chirality transfer from non-racemic solvents, such as (*S*)-1-chloro-2-methylbutane and (*R*)-2-chloroalkanes, to CD-silent poly(*n*-hexyl isocyanate) (PHIC) macromolecules was successfully realized (**Figure 2**). The generation of supramolecular chirality of PHIC polymers with a preferred handed helix was also demonstrated by CD signals. After the successful chirality transfer from solvent chirality to achiral polymers, the substrate structures with different functionality, mostly based on σ-conjugated polymers and π-conjugated polymers, are also investigated.

3.1. Preparation for optically active σ-conjugated polymers

Fujiki et al. found that a certain polysilane bearing remote chiral (*S*)-2-methylbutoxyphenyl groups was CD-silent when it was molecularly dispersed in tetrahydrofuran (THF) solution, due to the existence of dynamically equivalent amounts of right- and left-handed screw sense helical main chain domains. However, after addition of methanol, optically active polysilane aggregated in the good/poor cosolvents system and marked bisignate CD signals were abruptly observed in the UV region due to the Siσ-Siσ* transition [58]. This interesting observation intrigued them to investigate the chiral-solvent-induced aggregation processes of CD-silent polysilanes bearing achiral groups in tersolvents (chiral solvent/good solvent/poor solvent) systems (**Figure 3**). Then the clear exciton couplet CD signals of Siσ-Siσ* transitions of achiral polysilane aggregates with achiral *n*-propoxyphenyl and *n*-hexyl groups,

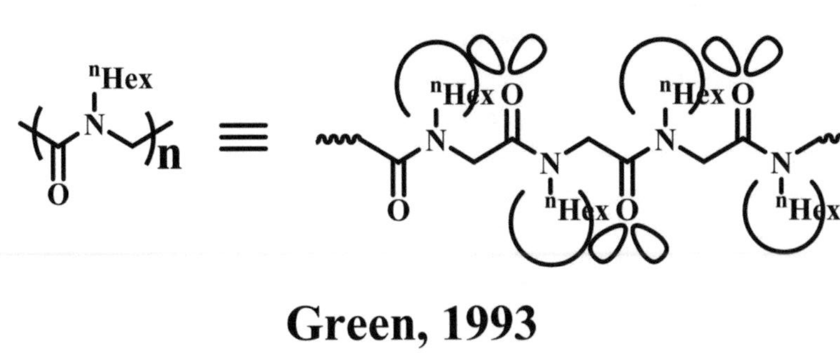

Figure 2. Conformational change of poly(*n*-hexyl isocyanate) driven by chiral solvation.

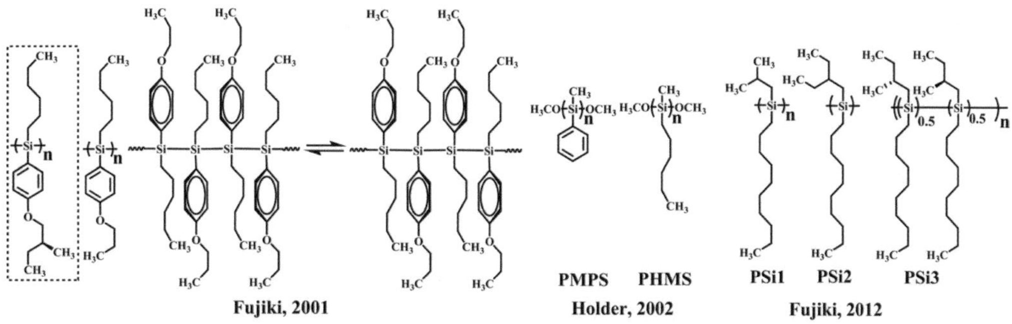

Figure 3. Optically activity of σ-conjugated polymers induced by chiral solvation.

certificated the successful chirality transfer and amplification of the molecular chirality of 2-phenylethanol (chiral solvent) [59]. Holder et al. first reported that preferential helical sense of inherently achiral σ-conjugated polysilanes was successfully induced via chiral solvation, which was investigated by optical UV-visible and CD spectroscopy [60, 61]. When polysilanes, poly(methylphenylsilane) (PMPS) and poly(methyl-*n*-hexylsilane) (PHMS) are totally dissolved in good solvents, they normally exist as random coils whose segments have an equal probability of adopting either *P* or *M* helical screw senses. While dissolved in chiral solvents, such as (*S*)-(−)-2-methyl-1-propoxybutane and (*S*)-(−)-(2-methylbutoxymethyl) benzene, a preference for one helical sense is achieved. Among these diverse chiral solvents mentioned above, limonene is most widely used as a mirror-symmetry-breaking solvent due to its nontoxicity and its easy extraction process from natural fruits. Through aggregation in the limonene/methanol/THF tersolvent, Fujiki et al. also successfully prepared optically active supramolecules from the original CD-silent σ-conjugated polysilanes (**Figure 3**) [62]. The chiral solvent ratio and polymer molecular weight dependences were both certified as critical factors to the dissymmetry factor, g_{CD}. Also, chirality inversion was observed when different limonene solvent ratios were applied to PSi1 and PSi3. And apparent CPL signals for PSi2 and PSi3 were observed in the supramolecular aggregates formed through chiral solvation. These interesting phenomena including chirality inversion and CPL signals are all instructive for research on solvent-chirality-transfer mechanism.

3.2. Preparation for optically active π-conjugated polymer

π-Conjugated polymers are very essential materials for applications in organic solar cells, polymeric organic light emitting diodes (OLED), thin-film transistors, lasers and photovoltaic devices. Among this, those polymers with chiroptical properties bring about perfect linearly and circularly polarized electroluminescence, which are widely applied in fields of optical switching and processing, chiral bio-imaging and metamaterials [63, 64]. As a typical π-conjugated polymer, achiral polyphenylacetylene possesses a plenty of short random twist segments with many helix-reversal points in the adjacent double bonds around a single bond. Yashima et al. reported that the random twist conformation can be transformed into a prevailing one-handed helical conformation upon complexation with optically active amines and amino alcohols in polar DMSO, certificated by the observation of characteristic ICD effects in the UV-vis region

(**Figure 4**) [53, 65, 66]. The CD signs corresponding to the different helical superstructures can be used as a probe for determining the configuration of chiral amines. What's more, this helicity can be successfully memorized when the original chiral amines are removed or replaced by achiral ones, while the helical conformation is still unchanged.

Among those π-conjugated polymers, polyfluorenes (PFs) possessing higher photoluminescence efficiency are much easier for film-processing and structure modification, which are great promising materials for organic light emitting diodes (OLED). Traditional methods for synthesizing PFs always involve expensive chiral catalysts and complex experimental procedures. The perfect application of chiral solvation method in preparing optically active σ-conjugated polymers and π-conjugated polyphenylacetylenes gave rise to the possibility of inducing optically active PFs. First, Fujiki et al. successfully utilized terpene chirality transfer method to generate serials of CD- and/or CPL-active π-conjugative polymers from the corresponding achiral counterparts (**Figure 5**) [67–71]. Meanwhile, many diverse factors, such as tersolvent composition, solvent polarity, polymer molecular weight, alkyl chain length, limonene enantiopurity, solution temperature, clockwise and counter-clockwise stirring, and aggregate size are confirmed to influence the magnitude of the induced CD and/or CPL amplitude. Through theoretical calculation, they assume that the inherent twisting ability (H-H repulsion) between the near-

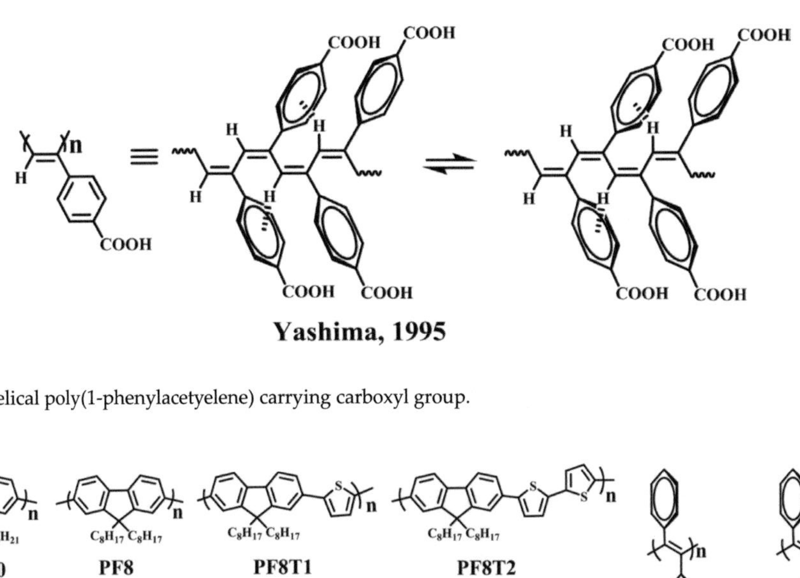

Figure 4. Helical poly(1-phenylacetyelene) carrying carboxyl group.

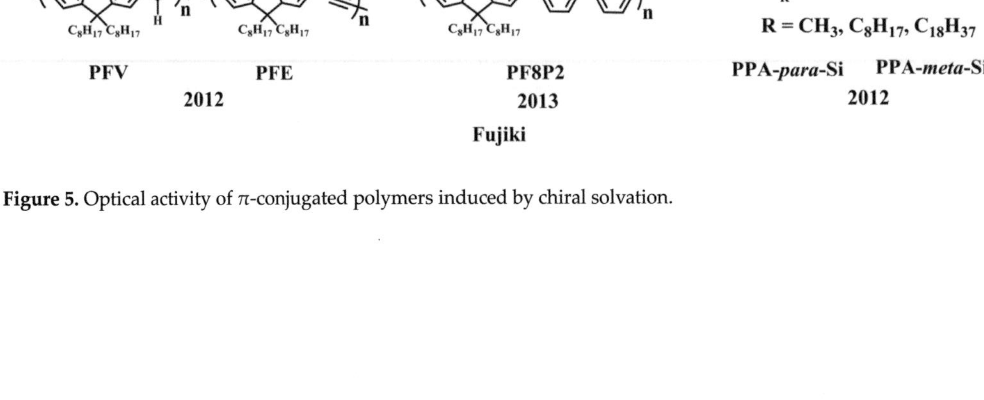

Figure 5. Optical activity of π-conjugated polymers induced by chiral solvation.

est neighbouring fluorene rings within mainchains of PF8T1, PF8T2, PF8, PF10 and PF8P2 are responsible for the preferential handness formed during aggregation process. In comparison, fluorene units linked with a C≡C triple bond (PEE) did not generate any optically active aggregate, ascribed to the lack of H-H repulsion.

Zhang et al. demonstrated that solvent chirality can be transferred to the aggregates of many optically inactive π-conjugated polymers with different backbone structures, such as main chain azo-containing polyfluorene (F8AZO), poly(9,9-di-n-octylfluorenyl-2,7-diyl) (PF8), poly(9,9-di-noctylsila-fluorenyl-2,7-diyl) (PSi8), poly(9-(1-octylnonyl)-9H-carbazole-2,7-diyl) (PCz8), P(F8-*alt*-Si8), P(F8-*alt*-Cz8), and P(Si8-*alt*-Cz8), hyperbranched PF8s, side chain azo-containing polymers PAzoMA, P(AzoMA-*rans*-MMA), and star side-chain Azo polymers (star PAzoMAs) as shown in **Figure 6** [69, 72–76]. Optically active F8AZO aggregates were successfully generated by the chirality transfer from (*S*)- and (*R*)-limonene, demonstrated by the intense ICD signals corresponding to F8AZO in the visible region [69]. Upon alternating photoirradiation at 405 (*trans* form) and 546 nm (*cis* form), a first reversible chiroptical switch for Azo-containing π-conjugated polymer was achieved, resulting from the switching between the *trans*-origin aggregation and *cis*-origin disaggregation of F8AZO in the limonene/2-propanol/chloroform tersolvent system. Then, they generated the first optically active hyperbranched π-conjugated polymer aggregates with strong CD and CPL properties with the help of solvent chirality transfer from chiral limonene [72]. Their studies showed that degree of branching, poor solvent type, volume fraction, limonene enantiopurity and polymer concentration have obvious effects on the magnitude and sign of the CD signals. Surprisingly, further studies found chiroptical inversion during aggregation of achiral PF8 and PSi8 and chiroptical inversion between CD and CPL spectra of PSi8 aggregates [73]. The unique chiroptical inversion was probably attributed to the opposite Mulliken charges between 9-Si in Si8 and 9-C in F8 unit and between C*ipso*(1) in Si8 and C*ipso*(1) in F8 unit, or resulting from the opposite direction of dipole moments in

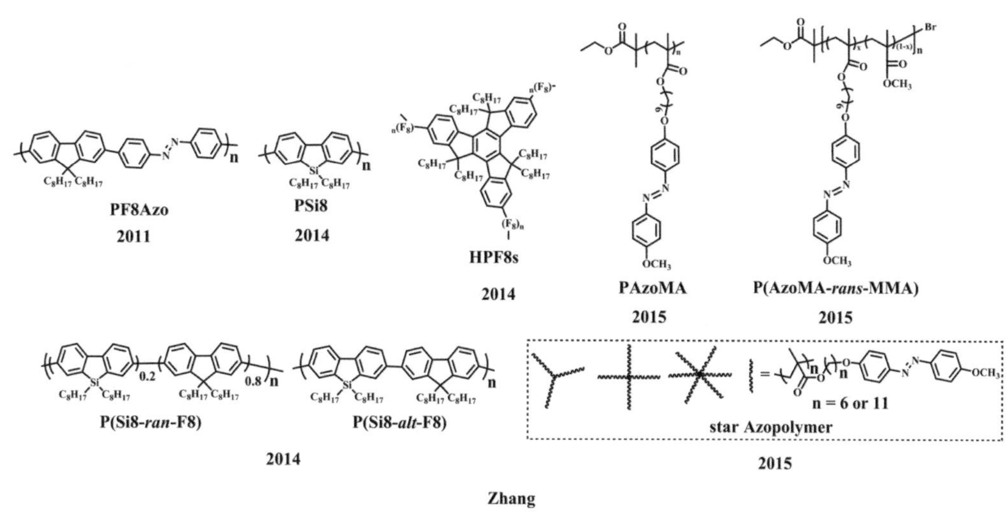

Figure 6. Optical activity of main-chain and side-chain π-conjugated polymers induced by chiral solvation.

three stable rotational isomers of the equatorial limonene rotamer. This novel element-dependent chiroptical inversion and structural dependence of π-conjugated polymers with the help of limonene chirality in aggregation states pave a new way for designing chiroptical functional polymers. Furthermore, they successfully constructed the supramolecular chirality for CD-silent PF8 by cooling its limonene solution at low temperature [74]. More interestingly, the supramolecular chirality can be transferred to solid films and be perfectly memorized. Facile generation and inversion of the CPL sign were also observed between aggregation states and film states. AFM studies clearly revealed the right- and left-handed twist helical fibres induced by nonracemic limonene, which are responsible for the CD and CPL functionality.

Although this chiral solvation strategy has been well applied in preparing many optically active main chain π- and σ-conjugated polymers, this concept has been extended to side chain polymeric systems. Guerra et al. reported that the first side chain polymer, syndiotactic polystyrene (s-PS), exhibited intense ICD signals in the polymer absorption region after exposure to nonracemic solvent vapour and thermal annealing processes in the film state [77]. The chiroptical properties obtained for racemic nanoporous δ-phase s-PS was assumed due to the induced chiral co-crystallization with solvent molecules, and the sign of CD signals is determined by the chirality of the nonracemic guests. However, a more recent study demonstrated that the signs depend essentially only on the nature of the polymer host supramolecular chirality, but not on the *R* or *S* solvent chirality, proven by the VCD spectra of the δ-phase s-PS film used [78]. To investigate the induction of supramolecular chirality for achiral side-chain polymers, Zhang et al. designed serials of linear and star-shaped achiral side chain azobenzene-containing polymer (PAzoMA, P(AzoMA-*rans*-MMA), and star PAzoMAs) [75, 76]. Chirality of nonracemic solvents were successfully transferred to these polymers, and well-assembled supramolecular *trans*-azobenzene aggregates are prepared in DCE/(*R*)- or (*S*)-limonene mixsolvents. The chirality obtained was ascribed to the ordered stacking of azobenzene group on the side chain when the polymer molecules were aggregating in the chiral solvents. This supramolecular chirality can also be destroyed by the *trans–cis* photoisomerization process due to the noncoplanar structure of the *cis*-Azo unit; however, it can be recovered by the heating-assisted reorganization process. The successful construction of a reversible chiral-achiral switch based on an achiral azobenzene-containing side chain polymer will open a new approach for production of chiroptical materials.

4. Conclusion

This chiral solvation approach provided herein allows the production of various CPL-/CD-functionalized polymer solutions or aggregates from CD-silent artificial polymers under mild conditions through the noncovalent interactions between small chiral molecules and achiral polymers. What's more, the chiral transfer, amplification and memory of supramolecular chirality in polymer solution and its solid polymer films were also realized, which is highly significant for the practical applications in chiroptical switch and memory, optical data storage and detection of circularly polarized luminescence (CPL). Considering the potential application of chiral materials in nonlinear optical devices, this concept paves a more convenient way for designing and constructing chiral polymer materials.

Acknowledgements

The authors are grateful for the financial support from the National Natural Science Foundation of China (21374072 and 21574089), the Priority Academic Program Development (PAPD) of Jiangsu Higher Education Institutions and the Program of Innovative Research Team of Soochow University.

Author details

Wei Zhang*, Yin Zhao and Lu Yin

*Address all correspondence to: weizhang@suda.edu.cn

State and Local Joint Engineering Laboratory for Novel Functional Polymeric Materials, Jiangsu Key Laboratory of Advanced Functional Polymer Design and Application, College of Chemistry, Chemical Engineering and Materials Science, Soochow University, Suzhou, China

References

[1] Wald G. The origin of optical activity. Annals of the New York Academy of Sciences. 1957;**69**(2):352–368. doi:10.1111/j.1749-6632.1957.tb49671.x

[2] Thiemann W, Darge W. Experimental attempts for the study of the origin of optical activity on earth. In: Oró J, Miller SL, Ponnamperuma C, Young RS, editors. Cosmochemical evolution and the origins of life. Dordrecht: Springer; 1974, pp. 263–283. doi:10.1007/978-94-010-2239-2_22

[3] Mason SF. Origins of biomolecular handedness. Nature. 1984;**311**(5981):19–23. doi:10.1038/311019a0

[4] Goldanskii VI, Kuz'min VV. Spontaneous mirror symmetry breaking in nature and the origin of life. AIP Conference Proceedings. 1988;**180**(1):163–228. doi: 10.1063/1.37867

[5] Bonner WA. Origins of chiral homogeneity in nature. In: Eliel EL, Wilen SH, editors. Topics in stereochemistry. Hoboken, NJ: John Wiley & Sons, Inc.; 2007, pp. 1–96. doi:10.1002/9780470147276.ch1

[6] Orgel LE. Molecular replication. Nature. 1992;**358**(6383):203–209. doi:10.1038/358203a0

[7] Páilyi G, Zucchi C, Caglioti L, editors. Advances in biochirality, 1st ed. Amsterdam: Elsevier; 1999, 407 p

[8] Feringa BL, van Delden RA. Absolute asymmetric synthesis: the origin, control, and amplification of chirality. Angewandte Chemie International Edition. 1999;**38**(23):3418–3438. doi:10.1002/(SICI)1521-3773(19991203)38:23<3418::AID-ANIE3418>3.0.CO;2-V

[9] Avalos M, Babiano R, Cintas P, Jiméneza JL, Palaciosa JC. Chiral autocatalysis: where stereochemistry meets the origin of life. Chemical Communications. 2000;(11):887–892. doi:10.1039/A908300F

[10] Lough WJ, Wainer IW, editors. Chirality in natural and applied science, 1st ed. Oxford: Wiley-Blackwell; 2002. 336 p

[11] Quack M. How important is parity violation for molecular and biomolecular chirality?. Angewandte Chemie International Edition. 2002;**41**(24):4618–4630. doi:10.1002/anie.200290005

[12] Cline DB. On the physical origin of the homochirality of life. European Review. 2005;**13**(S2):49–59. doi:10.1017/S1062798705000657

[13] Soai K, Kawasaki T. Asymmetric autocatalysis with amplification of chirality. In: Soai K, editor. Amplification of chirality. Berlin and Heidelberg: Springer; 2008, pp. 1–33. doi:10.1007/128_2007_138

[14] Cintas P. The origin of chirality in the molecules of life: a revision from awareness to the current theories and perspectives of this unsolved problem. By Albert Guijarro and Miguel Yus. Angewandte Chemie International Edition. 2009;**48**(12):2079–2080. doi:10.1002/anie.200805910

[15] Podlech J. Origin of organic molecules and biomolecular homochirality. Cellular and Molecular Life Sciences CMLS. 2001;**58**(1):44–60. doi:10.1007/PL00000777

[16] Yashima E, Maeda K, Iida H, Furusho Y, Nagai K. Helical polymers: synthesis, structures, and functions. Chemical Reviews. 2009;**109**(11):6102–6211. doi:10.1021/cr900162q

[17] Reggelin M, Doerr S, Klussmann M, Schultz M, Holbach M. Helically chiral polymers: a class of ligands for asymmetric catalysis. Proceedings of the National Academy of Sciences of the United States of America. 2004;**101**(15):5461–5466. doi:10.1073/pnas.0307443101

[18] Reggelin M, Schultz M, Holbach M. Helical chiral polymers without additional stereogenic units: a new class of ligands in asymmetric catalysis. Angewandte Chemie International Edition. 2002;**41**(9):1614–1617. doi:10.1002/1521-3773(20020503)41:9<1614::AID-ANIE1614>3.0.CO;2–5

[19] Okamoto Y. Chiral polymers for resolution of enantiomers. Journal of Polymer Science Part A: Polymer Chemistry. 2009;**47**(7):1731–1739. doi:10.1002/pola.23215

[20] Yashima E, Maeda K, Nishimura T. Detection and amplification of chirality by helical polymers. Chemistry–A European Journal. 2004;**10**(1):42–51. doi:10.1002/chem.200305295

[21] Maeda K, Yashima E. Dynamic helical structures: detection and amplification of chirality. In: Crego-Calama M, Reinhoudt DN, editors. Supramolecular chirality. Berlin and Heidelberg: Springer; 2006, pp. 47–88. doi:10.1007/128_035

[22] Yashima E, Maeda K. Chirality-responsive helical polymers. Macromolecules. 2008;**41**(1):3–12. doi:10.1021/ma071453s

[23] Kauranen M, Verbiest T, Boutton C, Teerenstra MN, Clays K, Schouten AJ, Nolte RJM, Persoons A. Supramolecular second-order nonlinearity of polymers with orientationally correlated chromophores. Science. 1995;**270**(5238):966–969. doi:10.1126/science.270.5238.966

[24] Akigi K. Helical polyacetylene: asymmetric polymerization in a chiral liquid-crystal field. Chemical Reviews. 2009;**109**(11):5354–5401. doi:10.1021/cr900198k

[25] Balavoine G, Moradpour A, Kagan HB. Preparation of chiral compounds with high optical purity by irradiation with circularly polarized light, a model reaction for the prebiotic generation of optical activity. Journal of the American Chemical Society. 1974;**96**(16):5152–5158. doi:10.1021/ja00823a023

[26] Green MM, Khatri C, Peterson NC. A macromolecular conformational change driven by a minute chiral solvation energy. Journal of the American Chemical Society. 1993; **115**(11):4941–4942. doi:10.1021/ja00064a086

[27] Wang Y, Sakamoto T, Nakano T. Molecular chirality induction to an achiral π-conjugated polymer by circularly polarized light. Chemical Communications. 2012;**48**(13):1871–1873. doi:10.1039/C2CC17027B

[28] Fujiki M. Supramolecular chirality: solvent chirality transfer in molecular chemistry and polymer chemistry. Symmetry-Basel. 2014;**6**(3):677–703. doi:10.3390/sym6030677

[29] Lehn JM, editor. Supramolecular Chemistry: Concepts and Perspectives, 1st ed. Weinheim: Wiley-VCH Verlag GmbH & Co. KGaA; 2006, 271 p. doi:10.1002/3527607439

[30] Snatzke G. Circular dichroism and optical rotatory dispersion — principles and application to the investigation of the stereochemistry of natural products. Angewandte Chemie International Edition in English. 1968;**7**(1):14–25. doi:10.1002/anie.196800141

[31] Nafie LA, Diem M. Optical activity in vibrational transitions: vibrational circular dichroism and Raman optical activity. Accounts of Chemical Research. 1979;**12**(8):296–302. doi:10.1021/ar50140a005

[32] Nafie LA. Infrared and Raman vibrational optical activity: theoretical and experiment aspects. Annual Review of Physical Chemistry. 1997;**48**:357–386. doi:10.1146/annurev.physchem.48.1.357

[33] Riehl JP, Richardson FS. Circularly polarized luminescence spectroscopy. Chemical Reviews. 1986;**86**(1):1–16. doi:10.1021/cr00071a001

[34] Allenmark S. Induced circular dichroism by chiral molecular interaction. Chirality. 2003;**15**(5):409–422. doi:10.1002/chir.10220

[35] Kipping FS, Pope WJ. LXXXVIII.-Racemism and pseudoracemism. Journal of the Chemical Society, Transactions. 1897;**71**:989–1001. doi:10.1039/CT8977100989

[36] Kipping FS, Pope WJ. LXIII.-Enantiomorphism. Journal of the Chemical Society, Transactions. 1898;**73**:606–617. doi:10.1039/CT8987300606

[37] Kipping FS, Pope WJ. XIV.-The crystallisation of externally compensated mixtures. Journal of the Chemical Society, Transactions. 1909;**95**:103–108. doi:10.1039/CT9099500103

[38] Hatano M, editor. Induced circular dichroism in biopolymer-dye systems. Berlin and Heidelberg: Springer; 1986, 121 p. doi:10.1007/BFb0071112

[39] Mason SF, Norman BJ. Outer-sphere co-ordination and optical activity in transition-metal complexes. Chemical Communications (London). 1965;(15):335–336. doi:10.1039/C19650000335

[40] Bosnich B. Asymmetric syntheses, asymmetric transformations, and asymmetric inductions in an optically active solvent. Journal of the American Chemical Society. 1967;**89**(24):6143–6148. doi:10.1021/ja01000a025

[41] Bezrodnaya TV, Mel'nik VI, Puchkovskaya GA, Savranskii LI. Vibrational and electronic spectra of benzophenone in different phase states: Ab initio calculations and experiment. Journal of Structural Chemistry. 2006;**47**(1):194–199. doi:10.1007/s10947-006-0286-8

[42] Hayward LD, Totty RN. Induced optical rotation and circular dichroism of symmetric and racemic aliphatic carbonyl compounds. Journal of the Chemical Society D: Chemical Communications. 1969;(12):676–677. doi:10.1039/C29690000676

[43] Noack K. Circular dichroism induction in an optically inactive compound by intermolecular interaction with an optically active solvent. Helvetica Chimica Acta. 1969;**52**(8):2501–2507. doi:10.1002/hlca.19690520833

[44] Kikuchi Y, Kobayashi K, Aoyama Y. Molecular recognition. 18. Complexation of chiral glycols, steroidal polyols, and sugars with a multibenzenoid, achiral host as studied by induced circular dichroism spectroscopy: exciton chirality induction in resorcinol-aldehyde cyclotetramer and its use as a supramolecular probe for the assignments of stereochemistry of chiral guests. Journal of the American Chemical Society. 1992;**114**(4):1351–1358. doi:10.1021/ja00030a033

[45] Kobayashi K, Asakawa Y, Kikuchi Y, Toi H, Aoyama Y. CH-π interaction as an important driving force of host-guest complexation in apolar organic media. Binding of monools and acetylated compounds to resorcinol cyclic tetramer as studied by ^{1}H NMR and circular dichroism spectroscopy. Journal of the American Chemical Society. 1993;**115**(7):2648–2654. doi:10.1021/ja00060a013

[46] Huang X, Rickman BH, Borhan B, Berova N, Nakanishi K. Zinc porphyrin tweezer in host-guest complexation: determination of absolute configurations of diamines, amino acids, and amino alcohols by circular dichroism. Journal of the American Chemical Society. 1998;**120**(24):6185–6186. doi:10.1021/ja973539e

[47] Kurtán T, Nesnas N, Li Y, Huang X, Nakanishi K, Berova N. Chiral recognition by CD-sensitive dimeric zinc porphyrin host. 1. Chiroptical protocol for absolute configurational assignments of monoalcohols and primary monoamines. Journal of the American Chemical Society. 2001;**123**(25):5962–5973. doi:10.1021/ja010249w

[48] Kurtán T, Nesnas N, Li Y, Huang X, Nakanishi K, Berova N. Chiral recognition by CD-sensitive dimeric zinc porphyrin host. 2. Structural studies of host–guest complexes with chiral alcohol and monoamine conjugates. Journal of the American Chemical Society. 2001;**123**(25):5974–5982. doi:10.1021/ja010250v

[49] Borovkov VV, Lintuluoto JM, Fujiki M, Inoue Y. Temperature effect on supramolecular chirality induction in bis(zinc porphyrin). Journal of the American Chemical Society. 2000;**122**(18):4403–4407. doi:10.1021/ja9936971

[50] Borovkov VV, Lintuluoto JM, Inoue Y. Supramolecular chirogenesis in zinc porphyrins: mechanism, role of guest structure, and application for the absolute configuration determination. Journal of the American Chemical Society. 2001;**123**(13):2979–2989. doi:10.1021/ja0032982

[51] Borovkov VV, Hembury GA, Inoue Y. Origin, control, and application of supramolecular chirogenesis in bisporphyrin-based systems. Accounts of Chemical Research. 2004;**37**(7):449–459. doi:10.1021/ar0302437

[52] Aimi J, Oya K, Tsuda A, Aida T. Chiroptical sensing of asymmetric hydrocarbons using a homochiral supramolecular box from a bismetalloporphyrin rotamer. Angewandte Chemie International Edition. 2007;**46**(12):2031–2035. doi:10.1002/anie.200604330

[53] Palmans ARA, Vekemans J, Havinga, Meijer EW. Sergeants-and-soldiers principle in chiral columnar stacks of disc-shaped molecules with C_3 symmetry. Angewandte Chemie International Edition. 1997;**36**(23):2648–2651. doi:10.1002/anie.199726481

[54] George SJ, Tomović Z, Smulders MMJ, de Greef TFA, Leclére PELG, Meijer EW, Schenning APHJ. Helicity induction and amplification in an oligo(*p*-phenylenevinylene) assembly through hydrogen-bonded chiral acids. Angewandte Chemie International Edition. 2007;**46**(43):8206–8211. doi:10.1002/anie.200702730

[55] George SJ, Tomović Z, Schenning APHJ, Meijer EW. Insight into the chiral induction in supramolecular stacks through preferential chiral solvation. Chemical Communications. 2011;**47**(12):3451–3453. doi:10.1039/c0cc04617e

[56] Stepanenko V, Li X, Gershberg J, Würthner F. Evidence for kinetic nucleation in helical nanofiber formation directed by chiral solvent for a perylene bisimide organogelator. Chemistry – A European Journal. 2013;**19**(13):4176–4183. doi:10.1002/chem.201204146

[57] Green MM, Park JW, Sato T, Teramoto A, Lifson S, Selinger RLB, Selinger JV. The macromolecular route to chiral amplification. Angewandte Chemie International Edition. 1999;**38**(21):3138–3154. doi:10.1002/(SICI)1521-3773(19991102)38:21<3138::AID-ANIE3138>3.0.CO;2-C

[58] Nakashima H, Fujiki M, Koe JR, Motonaga M. Solvent and temperature effects on the chiral aggregation of poly(alkylarylsilane)s bearing remote chiral groups. Journal of the American Chemical Society. 2001;**123**(9):1963–1969. doi:10.1021/ja000869h

[59] Nakashima H, Koe JR, Torimitsu K, Fujiki M. Transfer and amplification of chiral molecular information to polysilylene aggregates. Journal of the American Chemical Society. 2001;**123**(20):4847–4848. doi:10.1021/ja010119n

[60] Dellaportas P, Jones RG, Holder SJ. Induction of preferential helical screw senses in optically inactive polysilanes via chiral solvation. Macromolecular Rapid Communications. 2002;**23**(2):99–103. doi:10.1002/1521-3927(20020101)23:2<99::aid-marc99>3.3.co;2-d

[61] Toyoda S, Fujiki M. Cooperative preferential helical ordering in poly(alkylarylsilylene) copolymers. Macromolecules. 2001;**34**(3):640–644. doi:10.1021/ma000948d

[62] Nakano Y, Ichiyanagi F, Naito M, Yang Y, Fujiki M. Chiroptical generation and inversion during the mirror-symmetry-breaking aggregation of dialkylpolysilanes due to limonene chirality. Chemical Communications. 2012;**48**(53):6636–6638. doi:10.1039/c2cc17845a

[63] Oda M, Nothofer HG, Lieser G, Scherf U, Meskers SCJ, Neher D. Circularly polarized electroluminescence from liquid-crystalline chiral polyfluorenes. Advanced Materials. 2000;**12**(5): 362–365. doi:10.1002/(SICI)1521-4095(200003)12:5<362::AID-ADMA362>3.0.CO;2-P

[64] Gilot J, Abbel R, Lakhwani G, Meijer EW, Schenning APHG, Meskers SCJ. Polymer photovoltaic cells sensitive to the circular polarization of light. Advanced Materials. 2010;**22**(20):E131–E134. doi:10.1002/adma.200903995

[65] Yashima E, Matsushima T, Okamoto Y. Poly((4-carboxyphenyl)acetylene) as a probe for chirality assignment of amines by circular dichroism. Journal of the American Chemical Society. 1995;**117**(46):11596–11597. doi:10.1021/ja00151a032

[66] Yashima E, Maeda K, Okamoto Y. Memory of macromolecular helicity assisted by interaction with achiral small molecules. Nature. 1999;**399**(6735):449–451. doi:10.1038/20900

[67] Kawagoe Y, Fujiki M, Nakano Y. Limonene magic: non-covalent molecular chirality transfer leading to ambidextrous circularly polarised luminescent π-conjugated polymers. New Journal of Chemistry. 2010;**34**(4):637–647. doi:10.1039/B9NJ00733D

[68] Nakano Y, Liu Y, Fujiki M. Ambidextrous circular dichroism and circularly polarised luminescence from poly(9,9-di-*n*-decylfluorene) by terpene chirality transfer. Polymer Chemistry. 2010;**1**(4):460–469. doi:10.1039/B9PY00288J

[69] Zhang W, Yoshida K, Fujiki M, Zhu X. Unpolarized-light-driven amplified chiroptical modulation between chiral aggregation and achiral disaggregation of an azobenzene-*alt*-fluorene copolymer in limonene. Macromolecules. 2011;**44**(3):5105–5111. doi:10.1021/ma2012128

[70] Fujiki M, Jalilah AJ, Suzuki N, Taguchi M, Zhang W, Abdellatif MM, Nomura K. Chiral optofluidics: gigantic circularly polarized light enhancement of all-*trans*-poly(9,9-di-*n*-octylfluorene-2,7-vinylene) during mirror-symmetry-breaking aggregation by optically tuning fluidic media. RSC Advances. 2012;**2**(16):6663–6671. doi:10.1039/C2RA20430D

[71] Lee D, Jin Y, Kim H, Suzuki N, Fujiki M, Sakaguchi T, Kim SK, Lee WE, Kwak G. Solvent-to-polymer chirality transfer in intramolecular stack structure. Macromolecules. 2012;**45**(13):5379–5386. doi:10.1021/ma300976r

[72] Liu J, Zhang J, Zhang S, Suzuki N, Fujiki M, Wang L, Li L, Zhang W, Zhou N, Zhu X. Chiroptical generation and amplification of hyperbranched [small pi]-conjugated polymers in aggregation states driven by limonene chirality. Polymer Chemistry. 2014; **5**(3):784–791. doi:10.1039/C3PY01037F

[73] Wang L, Suzuki N, Liu J, Matsuda T, Rahim NAA, Zhang W, Fujiki M, Zhang Z, Zhou N, Zhu X. Limonene induced chiroptical generation and inversion during aggregation of achiral polyfluorene analogs: structure-dependence and mechanism. Polymer Chemistry. 2014;**5**(20):5920–5927. doi:10.1039/c4py00865k

[74] Zhao Y, Abdul Rahim NA, Xia Y, Fujiki M, Song B, Zhang Z, Zhang W, Zhu X. Supramolecular chirality in achiral polyfluorene: chiral gelation, memory of chirality, and chiral sensing property. Macromolecules. 2016;**49**(9):3214–3221. doi:10.1021/acs.macromol.6b00376

[75] Jiang S, Zhao Y, Wang L, Yin Lu, Zhang Z, Zhu J, Zhang W, Zhu X. Photocontrollable induction of supramolecular chirality in achiral side chain Azo-containing polymers through preferential chiral solvation. Polymer Chemistry. 2015;**6**(23):4230–4239. doi:10.1039/C5PY00496A

[76] Yin L, Zhao Y, Jiang S, Wang L, Zhang Z, Zhu J, Zhang W, Zhu X. Preferential chiral solvation induced supramolecular chirality in optically inactive star Azo polymers: photocontrollability, chiral amplification and topological effects. Polymer Chemistry. 2015;**6**(39):7045–7052. doi:10.1039/C5PY01175B

[77] Buono AM, Immediata I, Rizzo P, Guerra G. Detection and memory of nonracemic molecules by a racemic host polymer film. Journal of the American Chemical Society. 2007;**129**(36):10992–10993. doi:10.1021/ja0732936

[78] Rizzo P, Lepera E, Guerra G. Enantiomeric guests with the same signs of chiral optical responses. Chemical Communications. 2014;**50**(60):8185–8188. doi:10.1039/C4CC02853H

Chapter 5

Supramolecular Assembly and Stimuli-Responsive Behavior of Multielement Hybrid Copolymers

Conghui Yuan, Yiting Xu, Birong Zeng, Weiang Luo, Guorong Chen, Jie Mao, Cheng Liu and Lizong Dai

Additional information is available at the end of the chapter

http://dx.doi.org/10.5772/67905

Abstract

Toward the organic polymer, hybrid elements can be defined as those beyond C, H, O, and N. Polymers comprising hybrid elements, such as Si, P, B, or metal ions have attracted great attention in the design of high performance or smart materials. Introduction of hybrid elements into a polymeric network may also lead to the formation of new intermolecular interactions, thus promote the self-organization of polymer chains to form controllable structures and morphologies. In this chapter, we introduce some of the recent important development in the design and self-assembly of hybrid amphiphilic copolymers. Specific attention was paid on the hybrid amphiphilic copolymers containing POSS, boronic acid, or boronate functional moieties. We introduce the design, synthesis, self-assembly behavior, and properties of these hybrid amphiphilic copolymers in detail. Also, the advantages and drawbacks of these polymers and their corresponding nanoassemblies are discussed.

Keywords: hybrid polymer, self-assembly, stimuli-response, nanostructure, functionality

1. Introduction

The development of copolymer self-assembly depends not only on the exploration of reliable driving forces but also on the creation of new building blocks. Conventional organic copolymers composed of C, H, O, and N elements generally self-assemble through Van der Waals interactions. The highly dynamic and weak strength features of this intermolecular interaction usually bring a great challenge to the controllable self-assembly of copolymers. Recently, introduction of hybrid elements (such as Si, B, P, and metal ions) onto the polymer chains has become a favorable pathway to achieve novel properties. Notably, hybrid elements are of high capability to change the interaction among polymer chains. For example, the stacking ability of polyhedral oligomeric silsesquioxanes (POSS) units has discovered to be an effective driven force for the self-assembly of polymer chains. Also, intermolecular coordination created by B-N dative bond and organic-metal chelating can organize polymer chains into ordered three-dimensional structures because of their bonding directionality.

In this chapter, we focus on some of the recent development in design, synthesis, property, and potential applications of the nanoassemblies derived from multielement amphiphilic copolymers, especially those containing POSS, boronic acid, or boronate moieties. As the smallest organic/inorganic hybrid unit, POSS has gained great attention due to its well-defined structure, high thermal stability, easy functionalization, and so on [1–4]. Synergy between the inorganic core and the organic side groups renders POSS with excellent solubility in many solvents and high compatibility with various polymeric matrices. Thus, POSS and POSS-based polymers have been widely used in the modification of polymer resin to construct high performance organic/inorganic hybrid materials [5]. Herein, we pay particular attention to the POSS-based amphiphilic copolymers and their self-assembly behavior in solutions. Boronic acid moiety has a high binding affinity to saccharide, glycol, and catechol derivatives, and the as-formed boronate group is cleavable in response to glucose and acidic pH. Therefore, amphiphilic copolymers containing boronic acid or boronate moieties are of particular interest in the fabrication of smart nanoassemblies.

2. Design, self-assembly of multielement hybrid copolymers

2.1. Self-assembly of POSS-based amphiphilic copolymers

POSS cages are generally hydrophobic and have a larger diameter ranging from 1 to 3 nm. Amphiphilicity can be achieved even only one POSS cage is incorporated onto a hydrophilic polymer chain. The aggregation and ordered stacking of POSS cages have powerful strength to promote the solution self-assembly of POSS-based amphiphiles. This is particularly important for the realization of kinetic and thermodynamic equilibrium of self-assembly behavior, thus leading to the formation of well-defined assemblies. Chemical modification of POSS cages and polymerization of POSS cages with carbon-carbon double bond are the most used methods to create POSS-based amphiphilic copolymers [6–8]. The former method can easily generate one POSS-capped amphiphiles, whereas the second method usually results in the incorporation of multiple POSS cages onto one polymer chain.

2.1.1. POSS end-capped amphiphiles

POSS cages can be designed with functional groups that are reactive to hydrophilic polymer chains thus creating POSS end-capped amphiphiles. Among numerous POSS derivatives, primary amine functionalized POSS is usually used to prepare POSS end-capped amphiphiles because of their easy synthesis and the highly reactive primary amine group. For example,

http://dx.doi.org/10.5772/67905

starting from aminopropylisooctyl-POSS, we have designed and synthesized a hybrid amphiphile that contains a hydrophobic POSS head and a hydrophilic PEG tail functionalized with a bidentate ligand through amidation and esterification reactions [9]. Self-assembly of this amphiphile in water solution affords uniform assemblies, and their morphologies can be adjusted by Zn^{2+}, therefore endowing the assemblies with unique metal ion sensitivity (**Figure 1**).

POSS cages attached with a primary amine group could also be incorporated onto the end of a hydrophilic polymer chain through atom transfer radical polymerization (ATRP). Li et al. [10] constructed a aminopropylisooctyl-POSS end-capped amphiphilic copolymer poly (2-(2-methoxyenthoxy) ethyl methacrylate)-co-oligo(ethylene glycol) methacrylate) (POSS-P(MEO_2MA-co-OEGMA)) using POSS-Br as ATRP initiator (**Figure 2**). This polymer could self-assemble into spherical micelles comprising POSS core and P(MEO_2MA-co-OEGMA) corona in aqueous solution (**Figure 3**). The LCST and CP values of the thermoresponsive

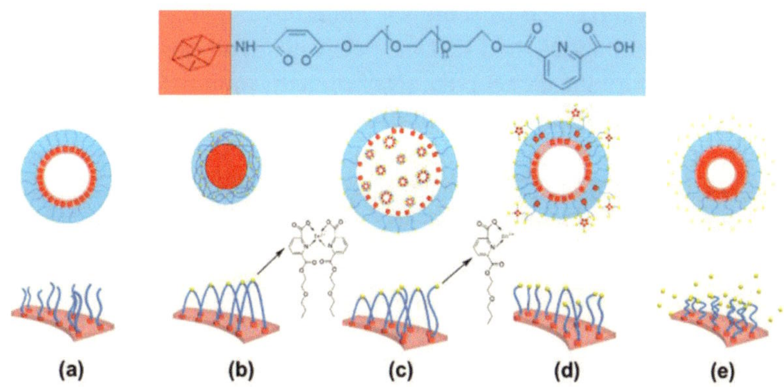

Figure 1. Proposed self-assembly model of Zn^{2+}/POSS-MA-PEG-DPA by metal coordination modulation.

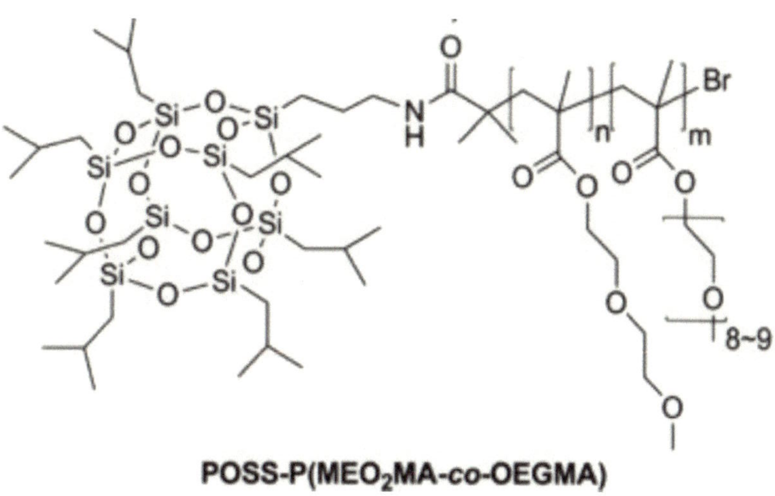

Figure 2. The chemical structure of POSS-P(MEO_2MA-co-OEGMA).

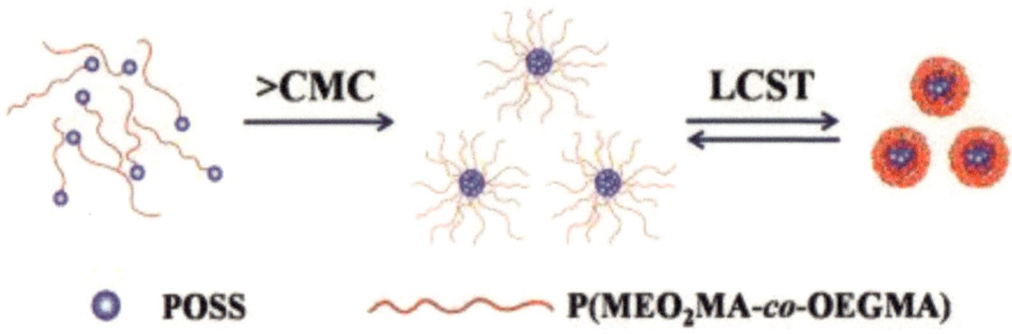

Figure 3. The schematic self-assembly process of the POSS-P(MEO_2MA-co-OEGMA) in water response to temperature.

polymers were increased with higher OEGMA content and could be well controlled by varying the ratio of MEO_2MA and OEGMA. Core-shell micelles and large aggregates were formed at temperatures below and above the LCST, respectively. Synthesis of aminopropylisooctyl-POSS end-capped amphiphilic POSS-poly (*N,N*-dimethylaminoethyl methacrylate)-block-poly (methyl methacrylate) (POSS-b-PDMAEMA-b-PMMA) (**Figure 4**) has also been reported by Wu et al. [11]. They found that POSS-b-PDMAEMA-b-PMMA could self-assemble into polymeric micelles with different shapes such as spherical, rod, and necklace morphology in different solutions. These micelles could encapsulate tetraphenylethene (TPE) and influence its fluorescent intensity. The introduction of POSS to polymer micelles could increase the fluorescence intensity of TPE, and spherical micelles could more tightly pack TPE to restrict intramolecular rotation of TPE to increase the fluorescent intensity compared with rod-like and necklace-like micelles. However, rod-like structure had a large specific surface area to interact with cell surface receptors, which enhanced its propensity to be internalized. Thus, the encapsulation of TPE aggregates in polymeric micelles not only facilitated the dispersion of TPE aggregates in biological environments but also enhanced the intracellular uptake of probes.

Yusa et al. [12] prepared a POSS end-capped amphiphiles using incompletely condensed (IC) and completely condensed (CC) POSS tethered with hydrophilic PEG chains (**Figure 5**). The association behavior of these amphiphiles in water solution was controlled based on

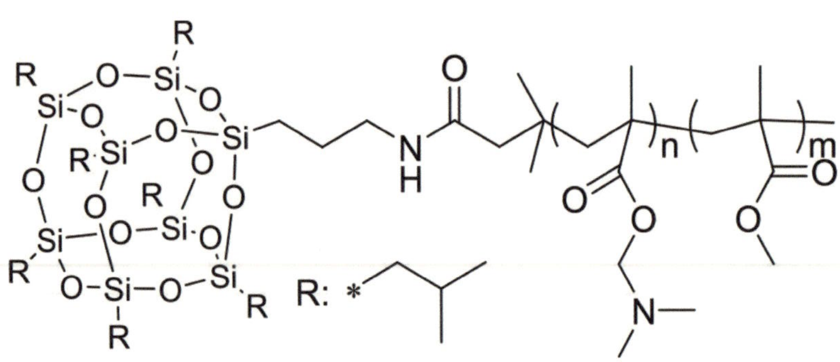

Figure 4. The chemical structure of POSS-b-PDMAEMA-b-PMMA.

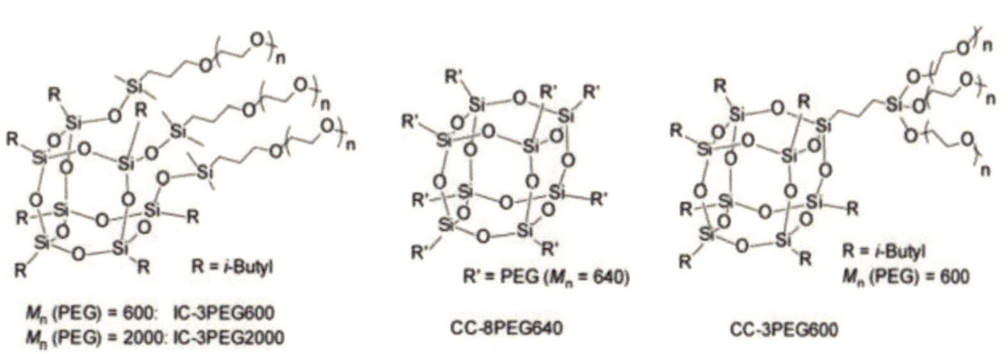

Figure 5. The chemical structures of amphiphilic POSS derivatives.

their chemical structure (**Figure 6**). The eight PEG chains-containing CC-POSS was hydrophilic and could be molecularly dissolved in water. Spherical micelles could be formed from IC-POSS in water solution as it contained three PEG chains with a molecular weight of 2000. Polydispersed worm-shaped micelles were formed by IC-POSS with three PEG chains of 600 molecular weight. Amphiphilic CC-POSS containing branched PEG chains with a molecular weight of 600 formed vesicle structures. These results indicated that the length of PEG chain and the shape of the POSS head group played a crucial role in determining the self-assembly structures.

2.1.2. *POSS containing amphiphilic block/random copolymers*

Copolymerization of carbon-carbon double bond bearing POSS cages with other water-soluble monomers was generally used to create amphiphilic block or random copolymers. Because of the larger diameter of POSS cage, the evident steric effect brings great difficult to the homo- and copolymerization of POSS monomers. This limitation was particularly evident in the chain extension of POSS macromolecular chain transfer agent during reversible

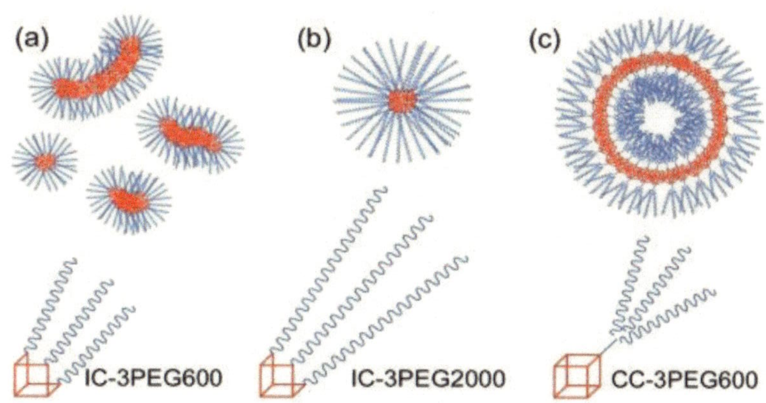

Figure 6. Schematic of the aggregates formed from (a) IC-3PEG600, (b) IC-3PEG2000, and (c) CC-3PEG600 in pure water.

addition-fragmentation chain transfer polymerization (RAFT). Therefore, great efforts have been focused on improving copolymerization activity of POSS monomers. For example, we have found that a trace amount of styrene (St) monomer can easily promote the chain extension of POSS macromolecular chain transfer agent for many monomers such as methyl methacrylate (MMA), acrylic acid (AA), 4-vinylpyridine (4VP), and *N*-isopropylacrylamide (NIPAm) [13]. This method is of great interest for the design and synthesis of POSS-based amphiphilic copolymers. Phase separation could be easily achieved during the self-assembly of POSS-based amphiphilic copolymers in water solution. We have synthesized a PMAPOSS-b-P(AA-co-St) block copolymer through RAFT polymerization and selective hydrolysis [14]. In a water solution, PMAPOSS-b-P(AA-co-St) with phase-separation patterns as core and hydrophilic segments as corona in aqueous media were obtained (**Figure 7**).

Xu et al. [15] reported a POSS-based hybrid pH-sensitive block copolymers poly(methacryl-isobutyl-POSS)-b-poly(4-vinylpyridine) (PMAiBuPOSS-b-P4VP) through RAFT polymerization. The size of aggregates in aqueous solution initially decreased and later increased as the pH value increased (**Figure 8**). It was supposed that this behavior was caused by the pH sensitivity of the P4VP block of the hybrid. Copolymerization of POSS monomers with more than one water-soluble monomers has also been demonstrated to be useful to create multicomponent hybrid amphiphilic copolymers. Xu et al. [16–18] also reported the synthesis of multicomponent organic/inorganic hybrid amphiphilic copolymers such as poly(methacrylate

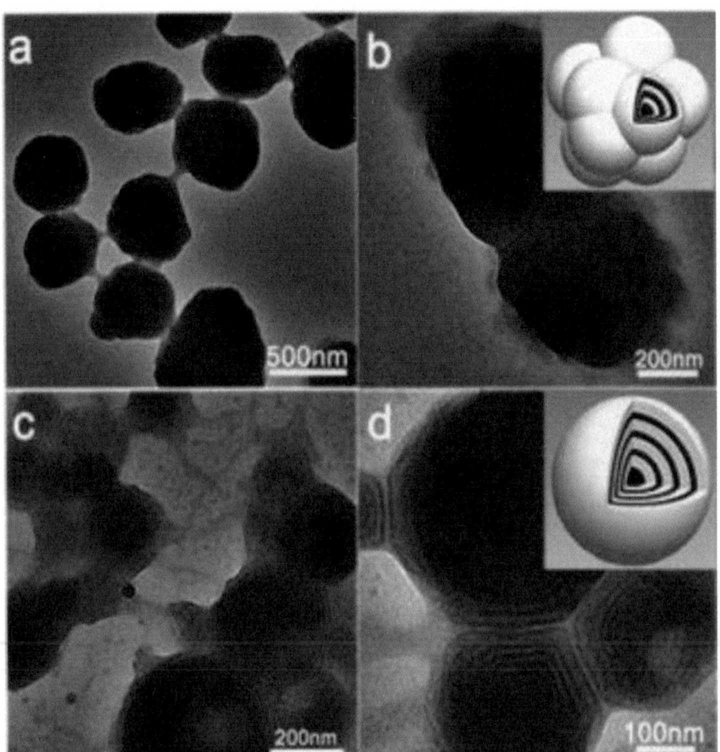

Figure 7. Morphologies of obtained BCPs aggregates in aqueous solutions.

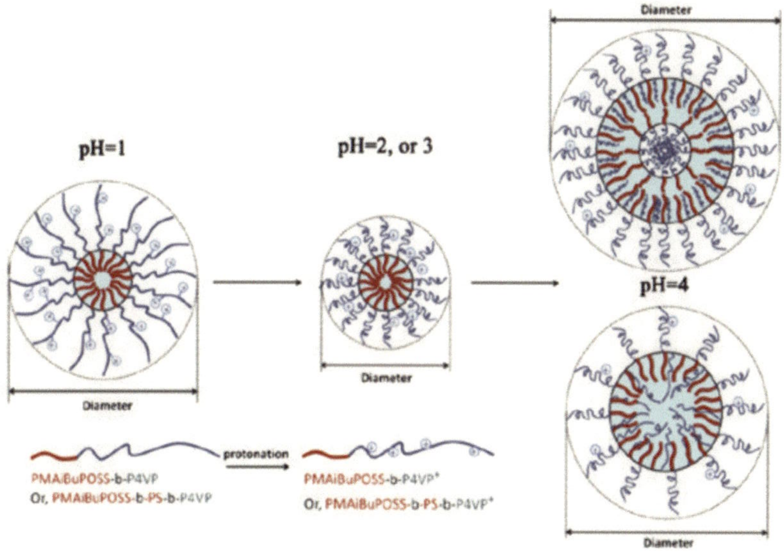

Figure 8. Schematic representation of the variation in aggregate size as a function of pH.

isobutyl POSS)-b-poly(*N*-isopropylacrylamide-co-oligo(ethylene glycol) methyl ether methacrylate) (PMAPOSS-b-P(NIPAM-co-OEGMA)) and poly(methacrylate isobutyl POSS-co-*N*-isopropylacrylamide-co-oligo(ethylene glycol) methyl ether methacrylate-co-2-vinylpyridine) P(MAPOSS-co-NIPAM-co-OEGMA-co-2VP) through RAFT polymerization. These random copolymers could self-assemble into spherical aggregates in water solution and show interesting stimuli-responsive behavior to pH, Zn^{2+}, or temperature (**Figures 9** and **10**). These hybrid assemblies could be potentially used in biological and medical fields, especially in drug nanocarriers for targeted therapy.

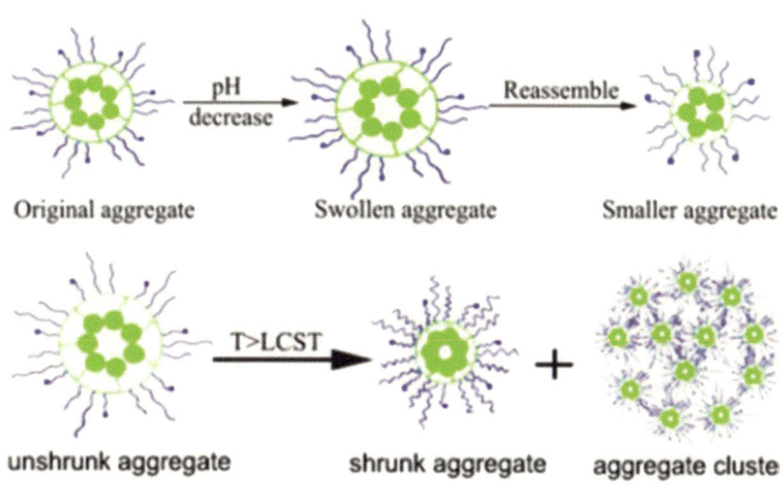

Figure 9. Schematic representation for the change in size of poly($MAPOSS_{2.5}$-co-$NIPAM_{80}$-co-$OEGMA_{202}VP_{40}$) aggregates responding to pH and temperature.

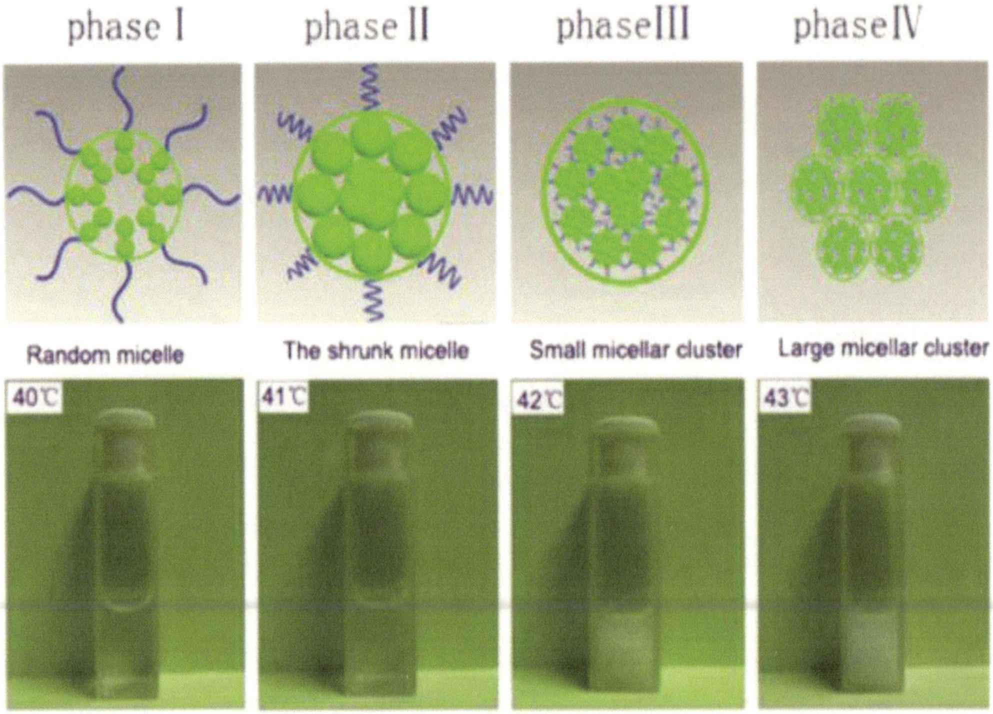

Figure 10. Schematic representation of the phase changes for PMAPOSS-b-P(NIPAM-co-OEGMA) micelle with the solution heating and the macroscopic phase transition of $PMAPOSS_9$-b-$P(NIPAM_{180}$-co-$OEGMA_{15})$ solution with the temperature ranging from 40 to 43°C.

Zeng et al. [19] synthesized a hybrid diblock copolymer consisting of bidentate ligand-functionalized chains by combining click reaction and RAFT polymerization. This copolymer was utilized to construct metal-containing polymer micelle by the metal-ligand coordination and electrostatic interaction. It was interesting to note that in common chloroform solvent, both the positively charged Zn^{2+} and negatively charged $AuCl_4^-$ could induce the formation of inverted micelles with a PVBT core and a PMAPOSS shell (**Figure 11**). Besides, the micelle could aggregate together to produce larger aggregates upon the addition of metal ions.

Matejka et al. [20] prepared a P(MMA-co-GMA)-b-PMAPOSS block copolymer by ATRP (**Figure 12**). In selective solvents, this polymer self-assembled to form ordered micellar-like structures. Spherical, cylindrical, or vesicle-like morphologies were produced by tuning the polymer and solvent composition. Cross-linking of the polymer by the reaction of the glycidyl group in the P(MMA-co-GMA) block of the micelle shell with a diamine results in a long-range structure ordering. The hexagonally packed cylindrical arrangement of the polymeric network was revealed by SAXS and TEM. The assemblies exhibited short-range structural order in solution and showed an order-disorder transition that is dependent on the solvent composition. Crosslinking the copolymers to form polymeric networks leads to stable long-range ordering.

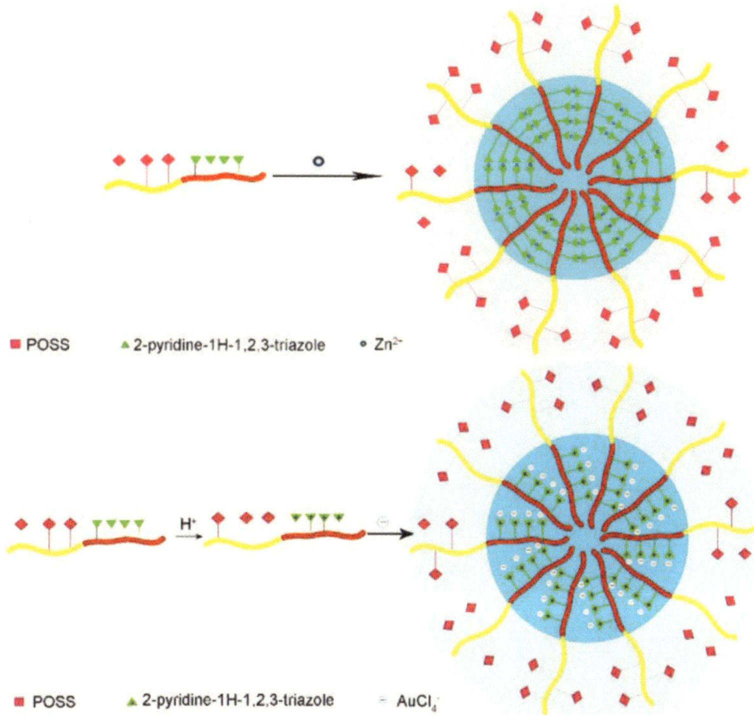

Figure 11. Schematic illustration of the Zn^{2+}-$PMAPOSS_{18}$-b-$PVBPT_{84}$ system and $HAuCl_4$-$PMAPOSS_{18}$-b-$PVBPT_{84}$ system.

Zhang et al. [21] reported a poly(maleimide isobutyl POSS-alt-vinylbenzyl polyethylene glycol) (P(MIPOSS-alt-VBPEG)) amphiphilic copolymer brushes with a sequence of alternating MIPOSS and PEG side chains through ordinary radical polymerization and RAFT polymerization (**Figure 13**). These alternating copolymer brushes had a low polydispersity index of less than 1.25. The DSC results showed that the crystallization behavior of PEG segments was greatly suppressed by the POSS moieties in copolymers. TGA results indicated that the thermal stability

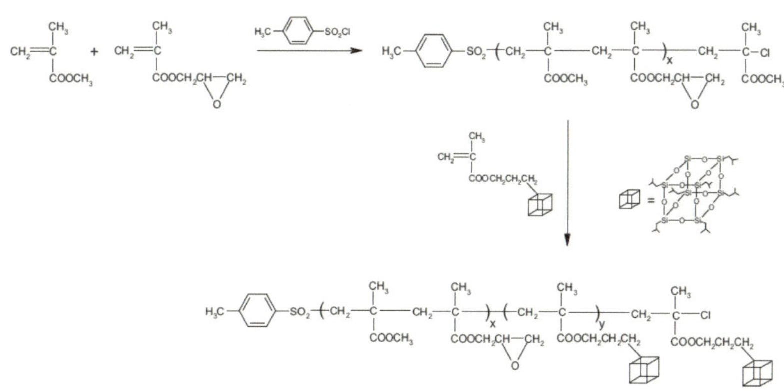

Figure 12. Synthesis of P(MMA-co-GMA)-b-PMAPOSS block copolymer.

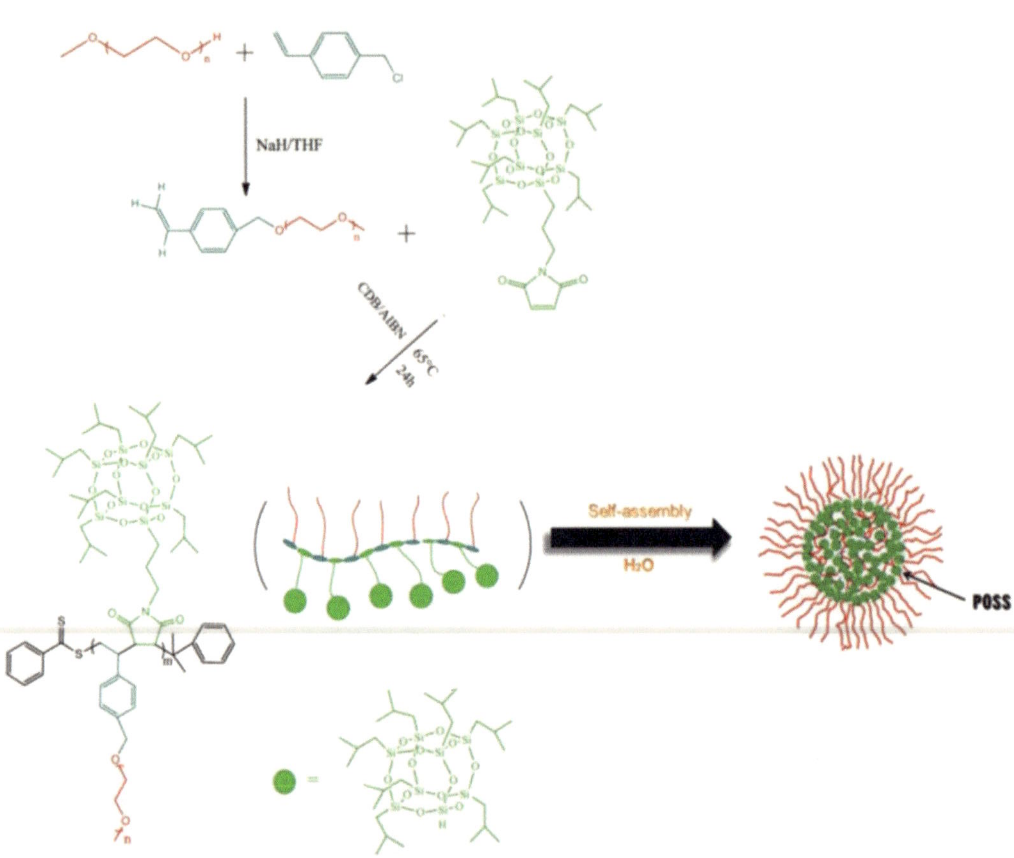

Figure 13. Synthetic route of P(MIPOSS-alt-VBPEG) and its self-assembly behavior in water solution.

of the P(MIPOSS-alt-VBPEG) could be enhanced by the incorporation of MIPOSS. These alternating copolymer brushes could form spherical aggregates in water. Moreover, the size of the aggregates increased on decreasing the chain length of the PEG monomer.

Hong et al. [22] also linked a POSS cage on a methacrylate monomer to afford a polymerizable POSS monomer (HEMAPOSS), which could efficiently decrease the steric hindrance of POSS cage in free-radical polymerization. Through a RAFT polymerization, PHEMAPOSS macro-chain transfer agent with a higher polymerization degree could be easily synthesized. Subsequently, *N,N*-dimethylaminoethyl methacrylate (DMAEMA) was adopted to achieve a chain extension of PHEMAPOSS to create a series of amphiphilic copolymers PHEMAPOSS-b-PDMAEMA. By simply varying the length of PDMAEMA block, these polymers could form assemblies with morphologies ranging from irregular aggregates, core-shell spheres, complex spheres (pearl-necklace-like structure) to large compound vesicles (**Figure 14**). Also, the assembly morphologies could transform reversibly from spherical micelles to complex micelles when cycling the solution from acidic to basic pH.

Wei et al. [23] reported a series of organic/inorganic random copolymers that were synthesized from methacrylate-terminated poly(ethylene oxide) (MAPEO) and 3-methacryloxypropylheptaphenyl POSS (MAPOSS) macromers through RAFT polymerization with

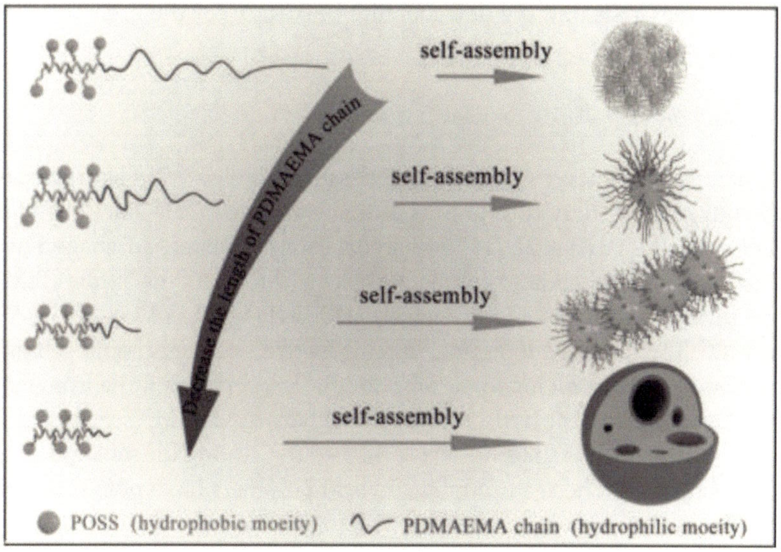

Figure 14. Self-assembly of PHEMAPOSS-b-PDMAEMA in water with the decreasing length of hydrophilic PDMAEMA chain.

4-cynao-4-(thiobenzoylthio) valeric acid as the chain transfer agent. The organic-inorganic random copolymers in bulk were microphase-separated and the POSS microdomains were formed through POSS-POSS interactions. In aqueous solutions, the organic-inorganic random copolymers were capable of self-assembling into spherical nanoobjects. The self-assembly behavior of the organic-inorganic random copolymers was also found to occur in the mixtures with the precursors of epoxy (**Figure 15**). The nanostructures were further fixed through subsequent curing reaction, and thus the organic-inorganic nanocomposites were obtained. In the organic-inorganic nanocomposites, the inorganic segments had a tendency to enrich at

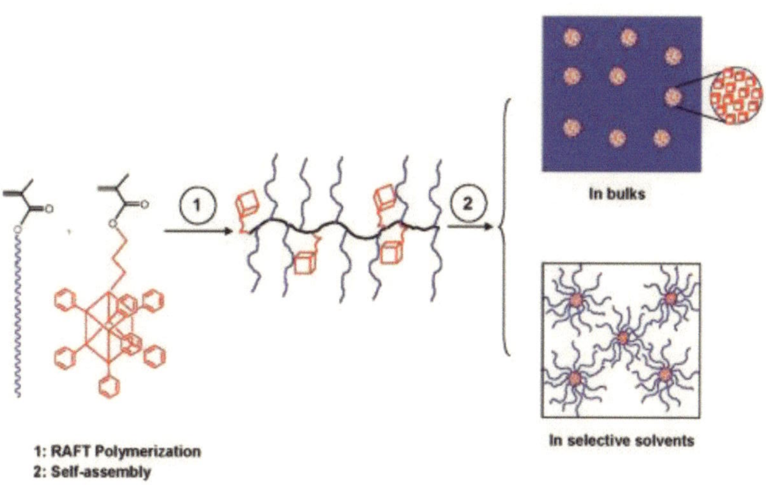

Figure 15. Self-assembly of P(MAPOSS-r-MAPEO) random copolymers.

the surface of the materials and the dewettability of surface for the organic-inorganic nanocomposites was improved.

2.1.3. *POSS-containing star-shaped amphiphilic polymers*

POSS cages generally have several organic moieties (at most eight). These organic moieties can be designed with reaction activity, and thus it can be used to create star-shaped amphiphilic polymers. For example, Yuan et al. [24] prepared a series of novel star-shaped hybrid P(2-(2-methoxyethoxy) ethylmethacrylate)-co-oligo(ethylene glycol) methacrylate (P(MEO$_2$MA-co-OEGMA)) polymers with a POSS core named (POSS-(P(MEO$_2$MA-co-OEGMA))$_8$) through ATRP (**Figure 17**). The obtained inorganic/organic hybrid polymers could self-assemble into micelles in aqueous solution owing to the amphiphilic property resulting from the hydrophobic inorganic POSS core and the hydrophilic P(MEO$_2$MA-co-OEGMA) segments. The tunable thermoresponse of the micelle solutions was achieved by varying the molar ratio of MEO$_2$MA and OEGMA (**Figure 16**). These amphiphilic hybrid polymers have potential applications in nanocarrier, nanoreactor, smart materials, and biomedical areas.

Li et al. [25] prepared a series of well-defined thermoresponsive amphiphilic star-shaped POSS-based inorganic/organic hybrid block copolymers of poly (ε-caprolactone)-poly(2-(2-methoxyethoxy)-ethyl methacrylate)-co-poly(ethylene glycol) methacrylate (POSS-(PCL-P(MEO$_2$MA-co-PEGMA))$_{16}$) through click chemistry, ring opening polymerization (ROP), and

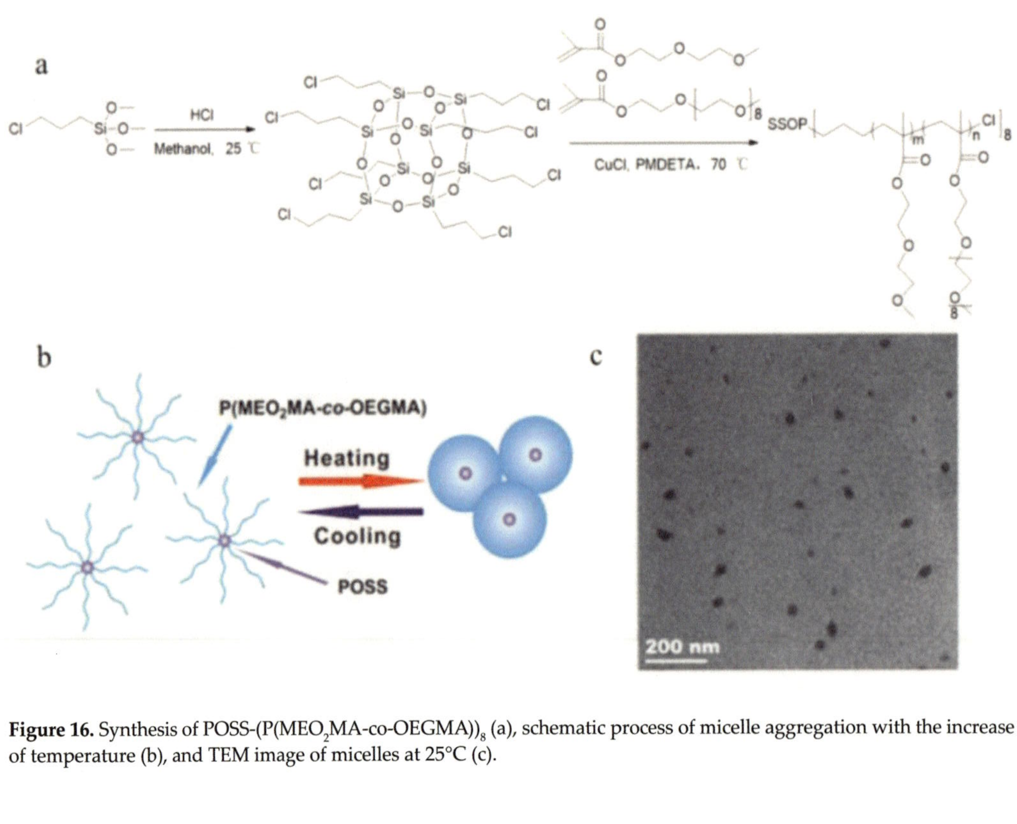

Figure 16. Synthesis of POSS-(P(MEO$_2$MA-co-OEGMA))$_8$ (a), schematic process of micelle aggregation with the increase of temperature (b), and TEM image of micelles at 25°C (c).

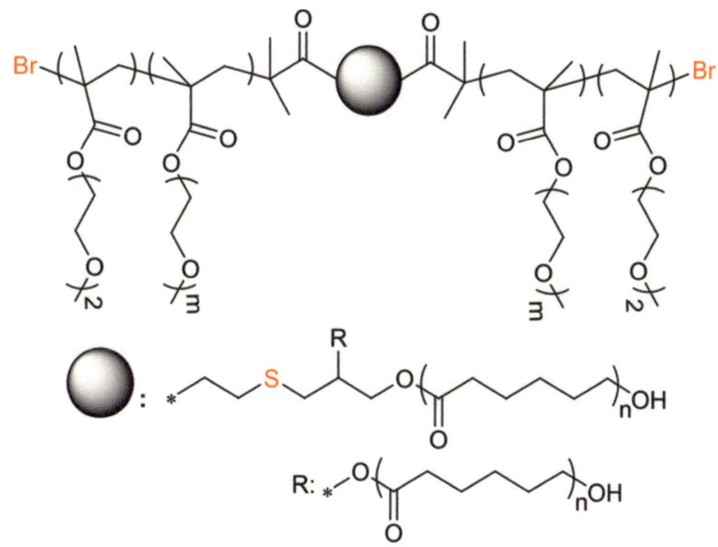

Figure 17. The chemical structure of POSS-(PCL-P(MEO_2MA-co-PEGMA)).

ATRP. By combining hydrophobic POS and PCL components with hydrophilic P(MEO_2MA-co-PEGMA) segments together, these copolymers could self-assemble into ellipsoidal aggregates with a moderately uniform size. Thermal-responsive behavior was observed to these organic/inorganic hybrid polymeric micelles. The critical phase transition temperature of these micelles in water solution could be finely tuned by changing the feed ratios of PEGMA and MEO_2MA. By increasing the content of PEGMA, the lowest critical solution temperature (LCST) of star-shaped POSS-(PCL-P(MEO_2MA-co-PEGMA))$_{16}$ increased from 34 to 57°C. At temperature higher than the corresponding LCSTs, the micelles aggregated to form spherical nanoparticles (**Figures 17** and **18**).

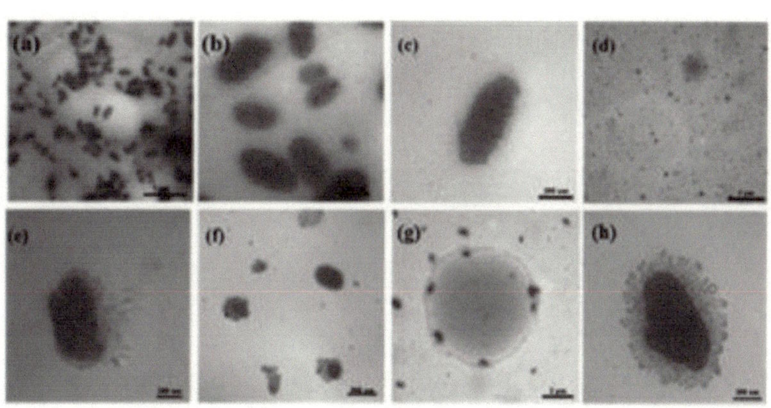

Figure 18. TEM images of POSS–(PCL–P(MEO_2MA-co-PEGMA))$_{16}$ micelles at 30°C (a–d) and at 50°C (e–h).

Li et al. [26] also prepared a well-defined pH-responsive amphiphilic POSS-containing star-shaped inorganic/organic hybrid block copolymers (BCPs), POSS-(PCL-P(MMAEMA-co-PEGMA))$_{16}$, through thiol-ene click reaction, ROP, and ATRP (**Figure 19**). These BCPs self-assembled into micelles in aqueous solution with a diameter about 175 nm. The stimuli-responsive behavior of these assemblies to solution pH and the controlled release of doxorubicin (DOX) were investigated. It was found that weakly acidic pH could cause the effective release of DOX up to 82 wt% (w/w). The low cytotoxicity, good biocompatibility, and excellent biodegradability make these micelles applicable in drug delivery. The DOX-loaded micelles could easily enter the cells and produce the desired pharmacological action, and minimize the side effect of free DOX (**Figure 20**).

2.2. Boron-containing amphiphilic copolymers

The development in organic chemistry have provided various routes to synthesize small molecules with boronic acid or boronate moieties, which are of great interest in the fabrication of polymers and functional materials containing boron element. There have been two general routes to create boronic acid or boronate polymers: (1) polymerization of boronic acid molecules with a carbon-carbon double bond; and (2) condensation reaction between multifunctional boronic acid molecules and catechol molecules. Because of the excellent response to biologies and pH, assemblies derived from boronic acid or boronate amphiphilic polymers have been widely used in biomedical applications.

2.2.1. Boronic acid-containing amphiphilic copolymers

Molecules with a boronic acid group attached on a benzene ring can be easily designed and synthesized. Carbon-carbon double bond can be further decorated onto the benzene ring to create polymerizable monomers. For example, Roy et al. [27] reported the synthesis of a low pKa boronic acid monomer, which was decorated with an electron withdrawing amide carbonyl on the phenylboronic acid moiety, and prepared an amphiphilic block copolymer

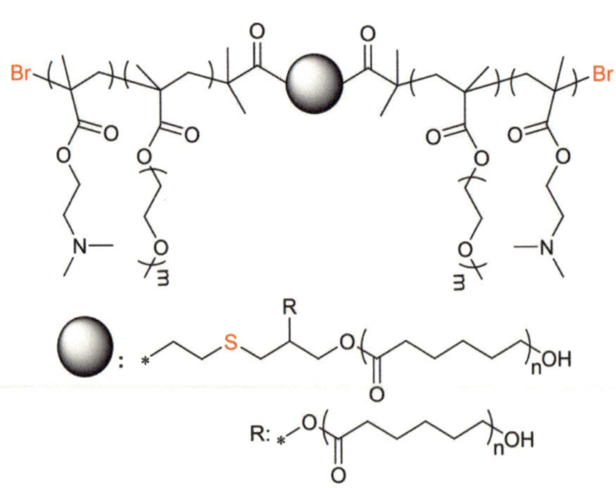

Figure 19. The chemical structure of POSS-(PCL-P(DMAEMA-co-PEGMA)).

http://dx.doi.org/10.5772/67905

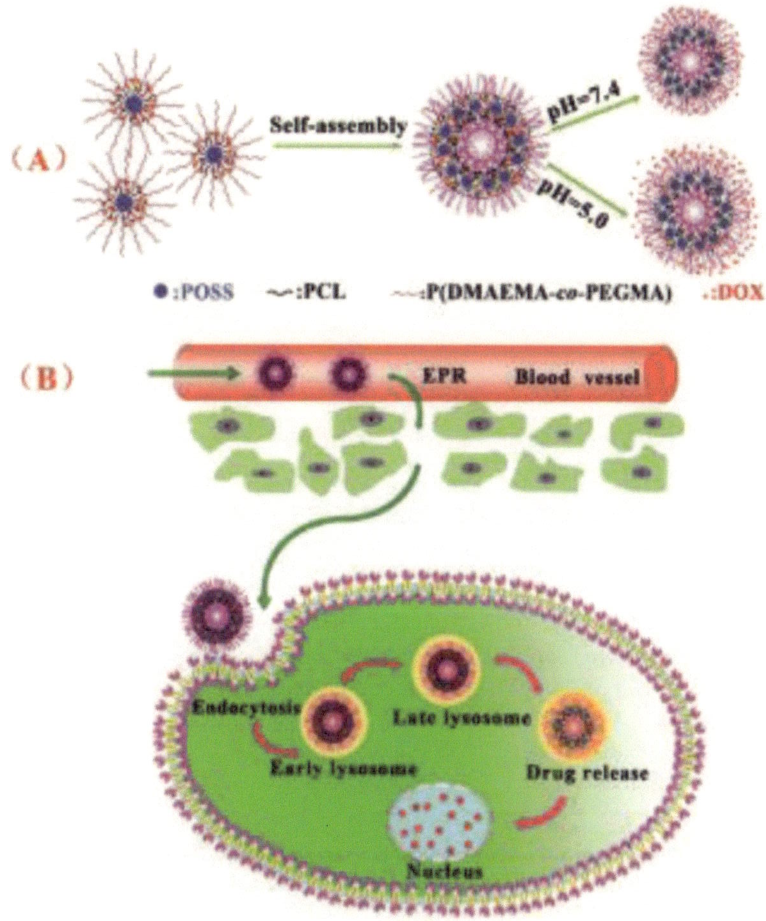

Figure 20. Illustration of pH-responsive self-assembly of the amphiphilic copolymer of POSS-(PCL-P(DMAEMA-co-PEGMA)) for the efficient intracellular release of anticancer drugs.

named PDMA-b-PAEBB via RAFT polymerization (**Figure 21**). This polymer could self-assemble into micelles composed of a hydrophilic PDMA corona and a hydrophobic PAEBB core. With the addition of sugar or the pH of aqueous media exceeding the pK_a of the block copolymers, the micelles dissociate, thus holding promise application in the areas of saccharide-sensing and self-regulated sugar-induced delivery.

Yuan et al. [28] developed a triply responsive amphiphilic copolymer by linking poly(ethylene oxide) (PEO) and poly(3-acrylamidophenylboronic acid) (PAPBA) with a disulfide bond. Micelles constructed with a PEO shell and a PAPBA core could be formed by the self-assembly of this copolymer. A simple adjustment of the solution pH led to the formation-dissociation of the micelles. Around the pKa of PAPBA segments, these micelles exhibited an interesting glucose response behavior by completely disassembling into copolymer solution. Moreover, these micelles could also dissociate through the breakage of disulfide bonds

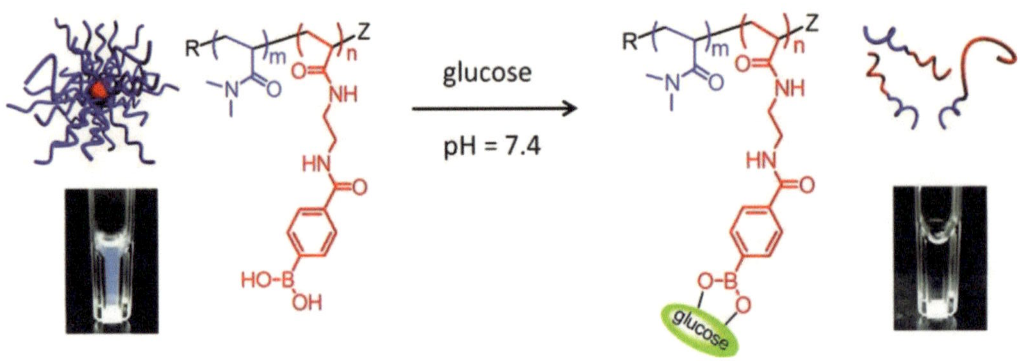

Figure 21. Schematic representation of the self-assembly of PDMA-b-PAEBB and their glucose-responsive behavior at pH = 7.4.

with the present of GSH. As a result, controllable release of insulin could be realized under the stimuli of glucose or GSH (**Figure 22**). Similar triply responsive amphiphilic copolymer assemblies were also described by Roy and coworkers [29]. They created a block copolymer through the RAFT copolymerization of a boronic acid acrylamido monomer and NIPAM and described the preliminary solution characterization of the resulting block copolymer. The boronic acid-containing copolymer displayed triply responsive behavior owing to the thermoresponsive nature of PNIPAM with the pH- and diol-responsive solubility of boronic acid-containing block.

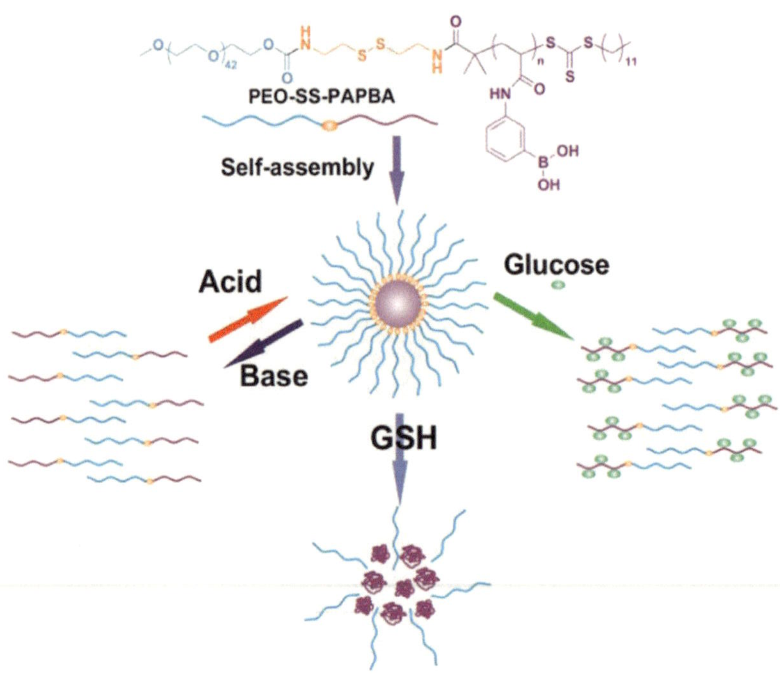

Figure 22. Self-assembly of the PEG-SS-PAPBA copolymer and glucose, pH, and redox triple responses.

Due to the reactivity of boronic acid group, monomers with a boronic acid moiety are hard to be polymerized through an ATRP method. Generally, the boronic acid group should be protected before ATRP. Kim and coworkers [30] synthesized a boronic acid polymer through ATRP route by employing a PEG macroinitiator to initiate polymerization with a pinacol-protected styrene boronic acid monomer. Then, deprotection of the pinacol-derived block copolymer afforded a boronic acid-containing PEG-PSBA block copolymer. Assemblies formed by this polymer were of great interest in smart reactors (**Figure 23**). Owing to the reversible hydrophilic-hydrophobic transition behavior of PSBA block, the use of stimuli-responsive block copolymers to construct nanoreactors with PSBA block as pore-generating components allowing the guest molecules to reach the enzyme residing inside the nanoreactors. In detail, they mixed PEG-PS and PEG-PSBA for the preparation of permeable polymersomes based on using an optimal amount of PEG-b-PSBA as a minor component; phase-separated PSBA domains would be dispersed throughout the PS matrix in the membrane of the polymersome, and these domains would be readily extracted from the matrix. They observed that when mixed in ratios of $W_{PSBA} = 30$ wt% and lower, the two block copolymers could coexist as a mixture in the membrane of the polymersome without experiencing complete phase separation, and the optimal mixing ratio is at ~ $W_{PSBA} = 10$ wt%. They also found that the polymersomes did not suffer from structural damages

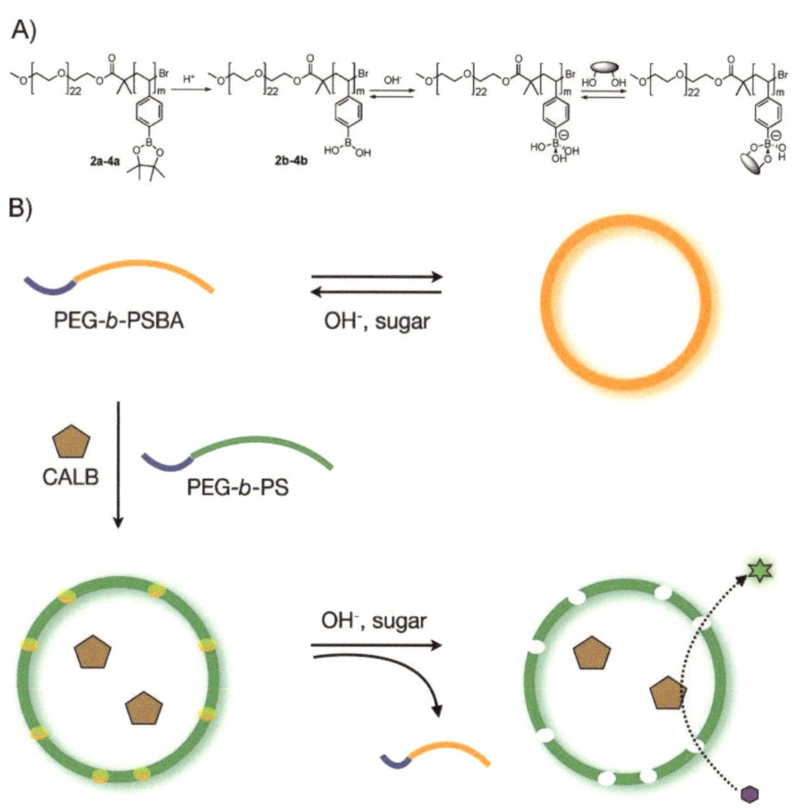

Figure 23. Synthesis of PEG-b-PSBA and schematic representation of the formation of bioreactors with a permeable membrane.

upon disassembly of the stimuli-responsive block copolymers observed by TEM. And the study of catalytic activity suggested that the differences in reactivity of bioreactors come from the difference in permeability of the polymersomes, which could be proportional to the area occupied by the phase-separated domains of PBSA block in the membrane.

To overcome the problem that boronic acid group brings to ATRP, thus acquiring boronic acid terminated polymers, Xu et al. [31] first synthesized alkyne-terminated PNIPAm using the BMP initiator and then conjugated the clickable boronic acid (APBA) with the obtained PNIPAm through the terminal alkyne via click reaction to get boronic acid terminated PNIPAm (**Figure 24**). The LCST of PNIPAm and BA-PNIPAm are 28.2 and 27.9°C, respectively. The CP of PNIPAm and BA-PNIPAm are 32.4 and 32.2°C, respectively. Owing to the excellent binding performance of BA-PNIPAm for saccharides and the LCST property of BA-PNIPAm, the author took advantage of the BA-PNIPAm to separate saccharides such as fructose from aqueous solution. Besides, they found that the fluorescence characteristic of APBA remained even after the terminal azide group is converted into atriazole after an alkyne-azide click reaction, and the intensity of fluorescence emission of BA-PNIPAm increased with the increasing fructose concentration because of the fluorescence emission of APBA was a result of possible B-N bond formed between the boron and one of the nitrogen atoms in the terminal azide.

Besides RAFT or ATRP routes, there are also some special methods to prepare amphiphilic copolymers with boronic acid groups. Aguirre-Chagala et al. [32] first reported a biodegradable derivatives of organoboron polymers (PPBC) synthesized by ring-opening polymerization (ROP) using a boronic acid-installed cyclic carbonates, catalyzed by 1,8-diazabicycloundec-7-ene (DBU) from a poly(ethylene glycol) macroinitiator (**Figure 25**). They used two types of monomers to synthesize a pinacol-protected or an acetonide-protected polymer, PPBC. And they found that deprotection of the acetonide-protected derivative could be achieved

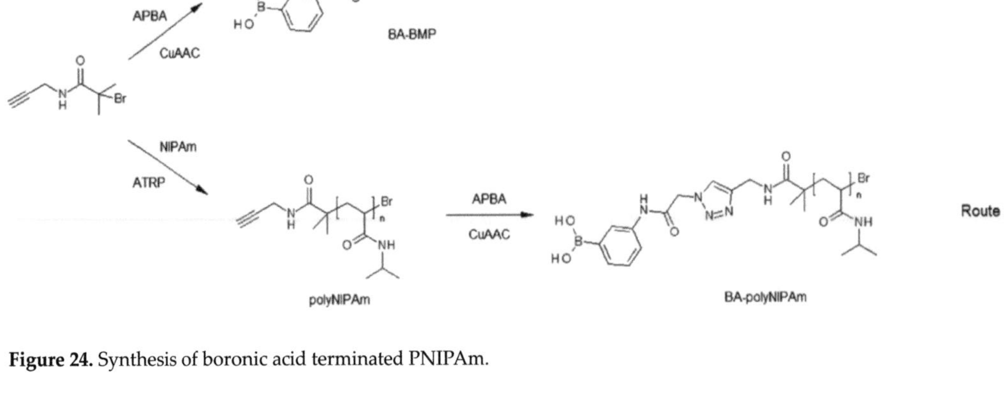

Figure 24. Synthesis of boronic acid terminated PNIPAm.

http://dx.doi.org/10.5772/67905

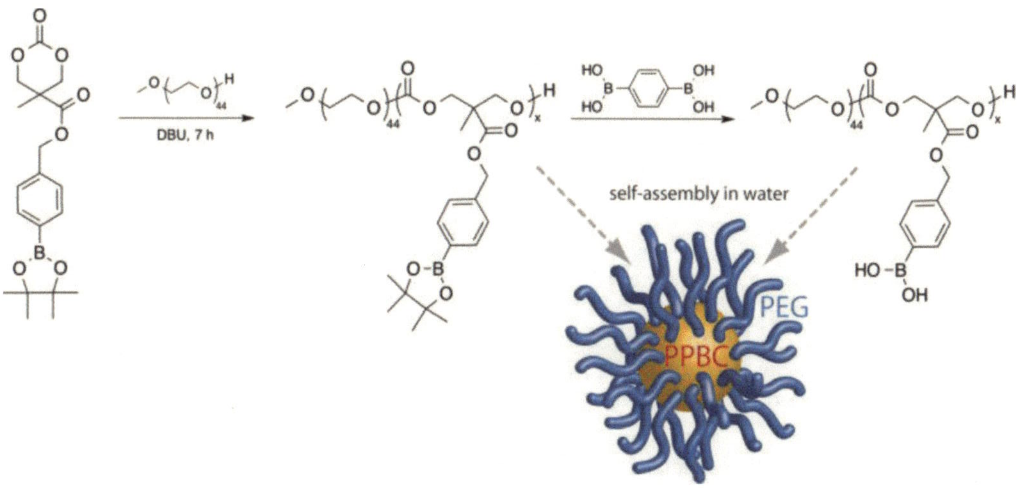

Figure 25. Synthesis and self-assembly of phenylboronic acid carbonate-based amphiphiles.

under more tractable conditions than the pinacol-protected polymer, by simply using an ion-exchange resin at room temperature. To study the self-assembly behavior of the PEG-b-PPBC copolymers, a multi-inlet vortex mixer (MIVM) was used to form aggregates by a rapid change in solvent quality. According to the results of TEM and static light scattering measurements, smaller, spherical aggregates were formed from amphiphiles with shorter PPBC lengths, while vesicle-like aggregates were observed for the copolymer with the longest PPBC block.

Since phenylboronic acid can form reversible covalent esters with 1,2- or 1,3-*cis*-diols including a ribose ring, small interfering RNA (siRNA) can be easily conjugated to the phenylboronic acid group. Naito and coworkers [33] synthesized a boronic acid-functionalized cationic polymer through modification of poly(ethylene glycol)-block-poly(L-lysine) (PEG-b-PLys) with 3-fluoro-4-carboxyphenylboronic acid (FPBA). They utilized the cationic polymer self-assembly into a micelle with siRNA as a cross-linker to stabilizing the polyion complex (PIC) and spontaneously delivering siRNA to targeted place, thus avoiding the leakage of the siRNA. They demonstrated that the PBA-assisted PIC micelles can be tailored to exhibit a dramatic disruption accompanied by the release of siRNAs in response to a change in the ribose concentration, which parallels events in the intracellular environment (**Figure 26**).

2.2.2. *Boronate-containing amphiphilic copolymers*

In comparison with boronic acid moiety, the boronate group is featured by its dynamic character, which is of great advantage in the construction of smart nanomaterials. Coumes et al. [34] designed a cleavable covalent block copolymer using boronate ester as a linker. They first synthesized two chain transfer agents terminated with a nitrodopamine or a boronic acid group, which were subjected to RAFT polymerizations with numerous monomers such as NIPAM, dimethylacrylamide (DMAc), n-butyl acrylate (nBuA), tert-butyl acrylate (tBuA), and St. Then, the coupling reaction between nitrocatechol- and boronic acid-terminated polymers was achieved via the nanoprecipitation method at room temperature, allowing the formation

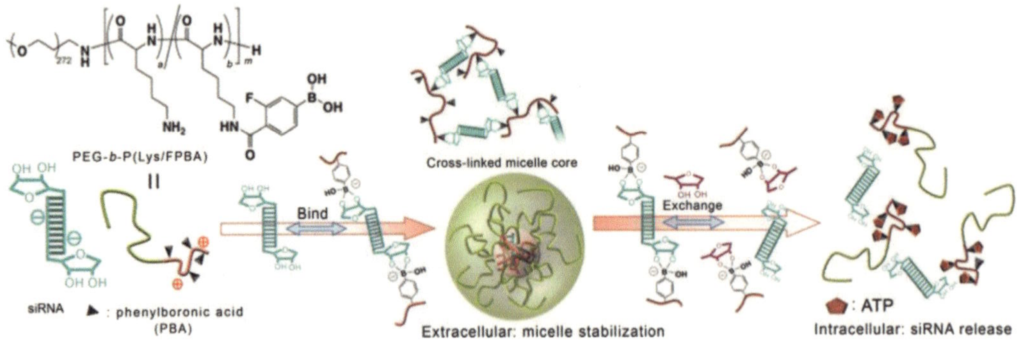

Figure 26. Schematic representation of the phenylboronic acid-based strategy for siRNA delivery.

of amphiphilic block copolymers. As the boronic esters are dynamic covalent structures, the amphiphilic copolymer can be cleaved at weakly acidic pH media or in the presence of sugars. Besides, the nitrocatechol derivatives can be chemical debonded by UV irradiation and cause the disassembly of the aggregates (**Figure 27**).

The formation of boronate could be also used to covalently attach functionality to the end of a polymer chain. De and coworkers [35] designed and synthesized a boronic acid-functionalized CTA by reacting 3-bromomethylphenylboronic acid with DMP and then prepared homopolymers and block copolymers with boronic acid functional end groups prepared by RAFT polymerization (**Figure 28**). They subsequently investigated their self-assembly behavior in aqueous or organic media.

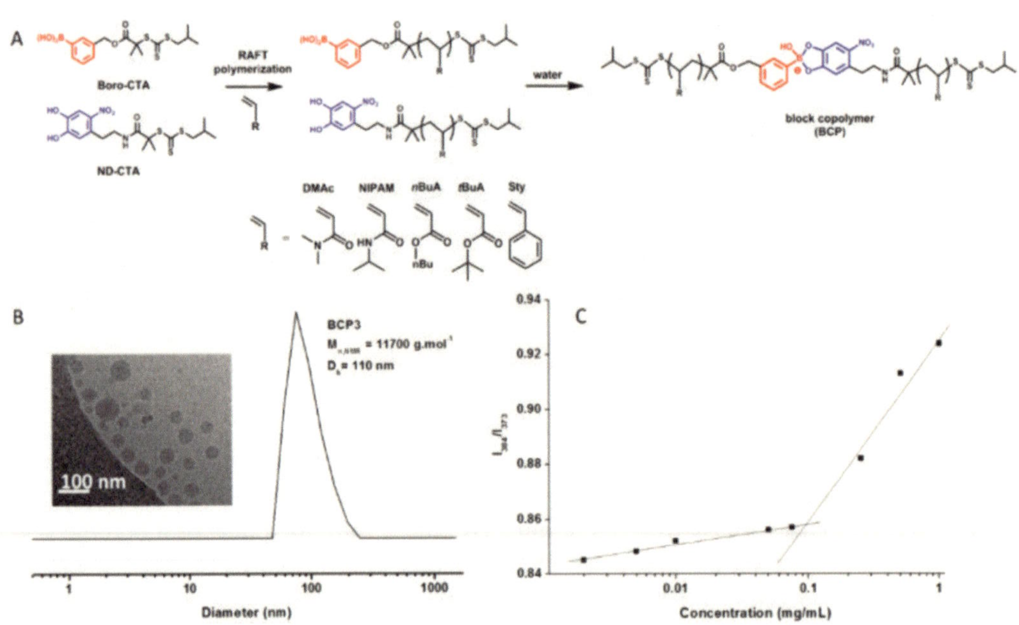

Figure 27. Synthesis and characterization of block copolymers (BCP) from nitrodopamine (ND-CTA)- and boronic acid (Boro-CTA)-functionalized CTAs.

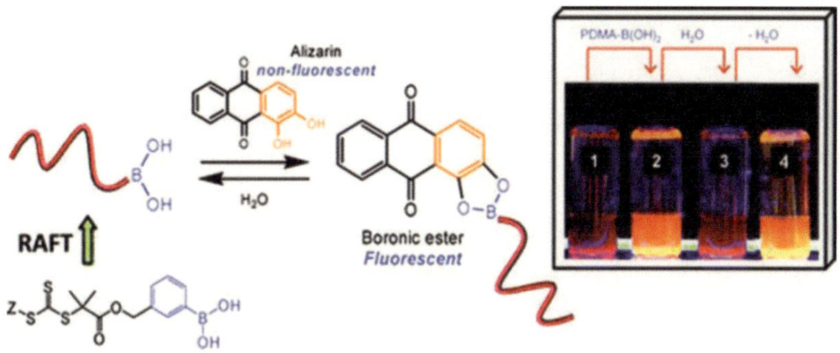

Figure 28. Proposed boronic ester formation between alizarin and boronic acid-terminated polymer.

It is well known that saccharide molecules having higher affinities with boronic acid groups than acyclic diols. The competition between saccharide molecules and acyclic diols to boronic acid groups may be useful in designing novel sugar-responsive materials. Yao and coworkers [36] designed and synthesized a new type of amphiphilic block copolymer with phenylboronate ester as a leaving group by ATRP. The block copolymer was amphiphilic because of the pinacol boronate ester units generated by the phenylboronic acid-acyclic diol complexation. With the presence of sugar molecules, boronic acid groups were prone to combine with sugar molecules; therefore, the pinacol phenylboronate moieties detaching from polymer and triggering the polarity transfer of the polymer from amphiphile to double hydrophile. As a result, the polymer nanoaggregates disassembled, thus causing the release of the loaded guest molecules (**Figure 29**).

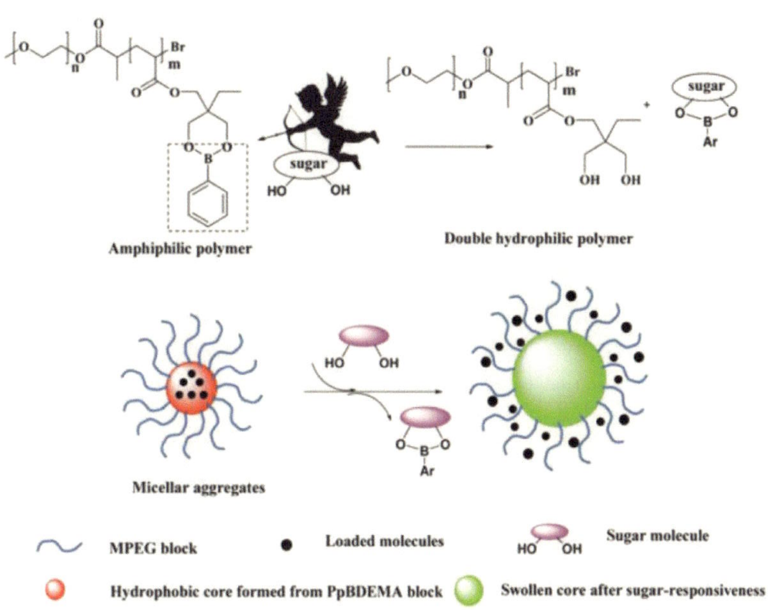

Figure 29. Schematic diagram for sugar-responsive behavior.

2.2.3. *Supramolecular assembly derived by dative B-N bond*

Boronate esters are Lewis acidic compounds, which can form B-N dative bond with N-donor ligands. Sheepwash and coworkers [37] utilized this B-N interaction for the creation of crystalline organic networks and organogel by connecting triboronate esters with bipyridyl linkers. They used two kinds of triboronic acid along with 4-tert-butylcatechol and 4,4′-bipyridine or 1,2-di(4-pyridyl)ethylene as the model molecules. The triboronic acid undergo a triple condensation reaction with the catechol to give a triboronate ester in organic solvents such as chloroform or toluene, which were then linked by the bipyridyl linker to form a polymer precipitated upon cooling. Polymers can be crystallized by slow cooling of toluene solutions or by vapor diffusion of pentane into toluene solutions. They also found that condensation reactions implemented between the extended triboronic acid, which was obtained from 1,3,5-tris(4′-bromobiphenyl)benzene, with 4-tert-butylcatechol and 4,4′-bipyridine in toluene resulted in the formation of an orange gel. The gel formation can also be reversed by increasing the temperature.

In order to investigate the factors that govern the binding strengths of dioxaboroles, Sheepwash and coworkers [38] utilized ^{1}H NMR spectroscopic titration experiments to study the binding constants of K_a between Lewis-acidic arylboronate esters and nitrogen-donor ligands with different electron effect substituents. The use of appropriate molecular components with both an electron withdrawing substituent on the dioxabole segment and an electron donating group on the pyridyl ligands has led to a significant improvement in the K_a values. This finding provided a new way to create main-chain supramolecular polymers with dative B-N bonds. Therefore, they designed and synthesized three monomers by condensation of the 4-(6-(methyl(pyridin-4-yl)amino)hexyl)benzene-1,2-diol with different arylboronic acids. Indeed, these monomers self-assembled into aggregates through the formation of B-N dative bond (**Figure 30**). This study turn out that main-chain supramolecular polymers could be achieved based on dative B-N bonds, and the B-N binding motif can also be used for the formation of branched or cross-linked supramolecular polymers.

Li et al. [39] have developed a cooperative polymerization strategy to prepare nanospheres using B-N dative bond as the driving force. Two monomers incorporated with boronic acid and catechol groups were prepared (**Figure 31**). When the methanol solutions of these two monomers were mixed together, boronate ester polymers were formed through sequential boronate esterification. Simultaneously, B-N dative bond formed between imine and boronate moieties in the polymer chains led to the formation of uniform nanospheres. It was found that the formation of the nanospheres adopted a N-E polymerization process, and the size of the nanospheres could be well controlled (**Figure 32**). This was the first example for the cooperative polymerization using B-N dative bond as a driving force.

We have also found that intermolecular B-N dative bond can be formed among small molecules [40]. A planar boronate ester (BCe), containing two imine moieties, a boronate ester group and two polymerizable carbon-carbon double bonds, has been created by the condensation reaction between a boronic acid molecule (BM) and a catechol molecule (CM) (**Figure 33**). In a methanol solution, B-N dative bond occurs among BCe molecules, thus leading to the formation assemblies with different morphologies. These assemblies can be further stabilized

http://dx.doi.org/10.5772/67905

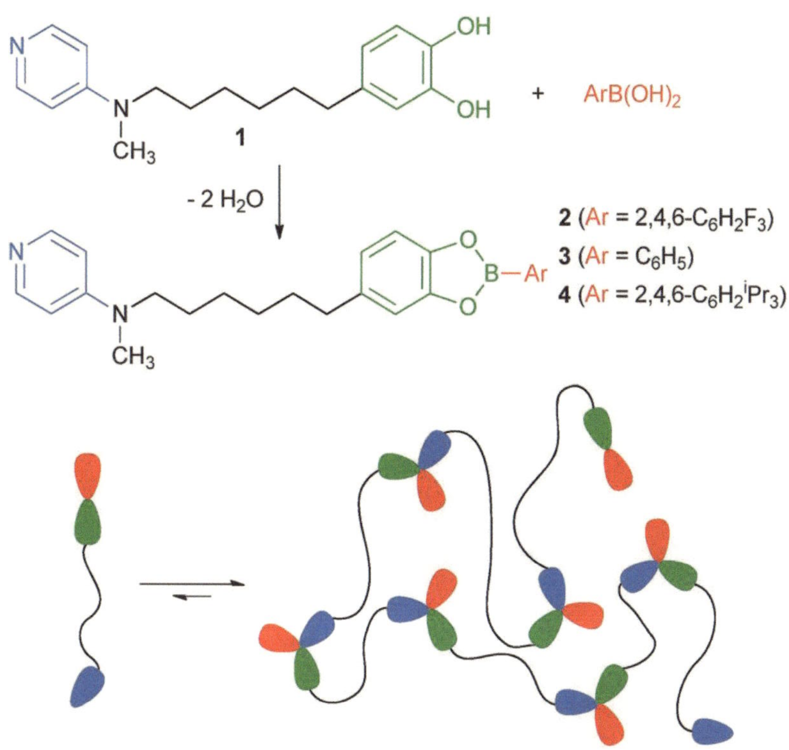

Figure 30. Synthesis of the heteroditopic monomers and aggregation through B–N dative bonds.

through the polymerization of carbon-carbon double bonds. Dramatic fluorescence intensity evolution has been observed along with the degradation of these assemblies under the stimuli of pH or D-glucose. Importantly, the reformation of B-N dative bond endows the assemblies with excellent self-healing property.

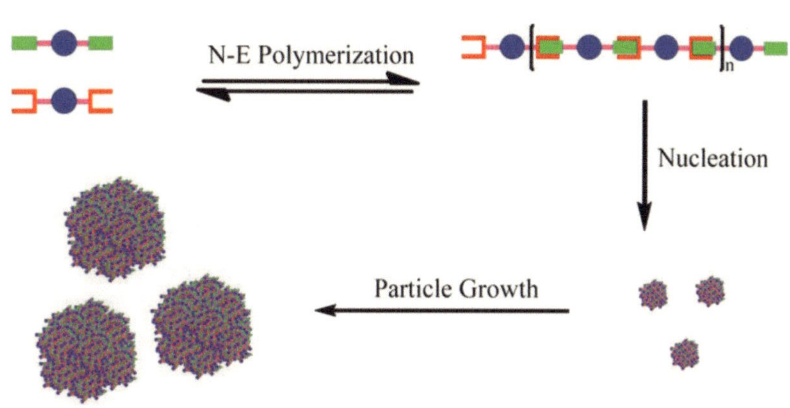

Figure 31. Schematic representation of the B-N dative bond derived boronate nanospheres.

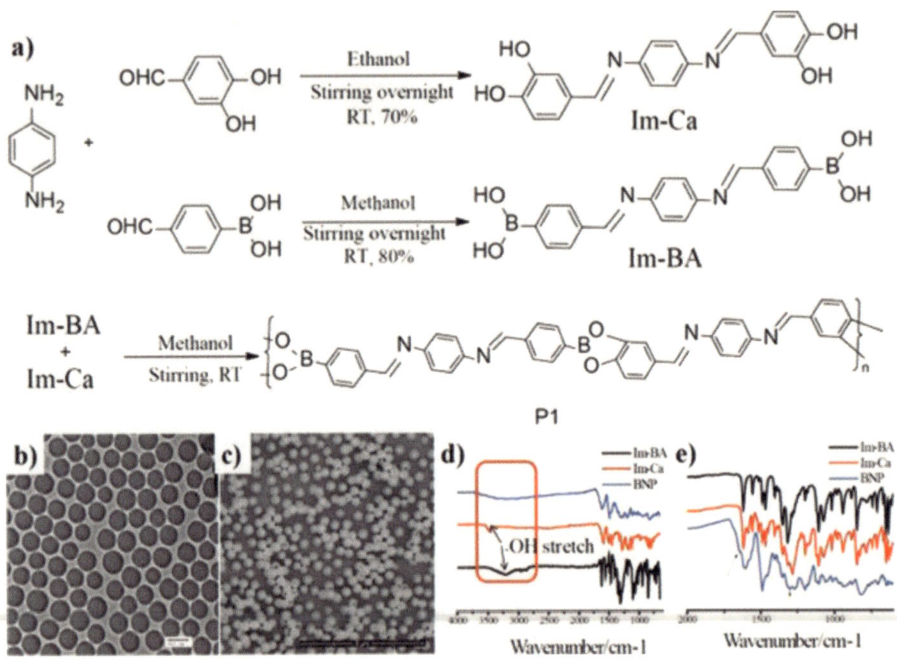

Figure 32. Synthesis and characterization of boronate nanospheres.

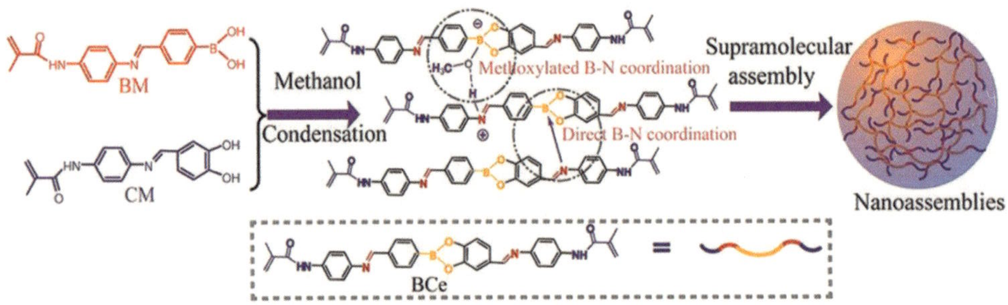

Figure 33. Synthesis and supramolecular assembly of BCe.

3. Conclusion

Polymer chemistry has provided numerous methods to design and synthesize polymers with various compositions and structures. This is the fundamental support for the explosion of high performance or functional materials. Up to now, polymers comprising only C, H, O, and N elements have been extensively investigated and the limitations of these materials have been well understood. Introduction of hybrid elements such as Si, P, As, B, or even metal ions into polymer chains have gained great interest in developing new polymeric materials with enhanced or new properties. In addition, new intermolecular interactions among polymer chains may emerge in polymers containing hybrid elements. This is particularly important for the self-assembly of polymers to form novel structures and morphologies.

http://dx.doi.org/10.5772/67905

However, great challenge has been encountered in the synthesis of hybrid polymers: (1) synthesis of monomers containing hybrid elements is usually difficult; (2) many hybrid monomers have a low polymerization activity; (3) purification and processing of hybrid polymers normally need complicated procedures. Fortunately, the development of organic and polymer chemistry has provides great opportunities for the design of new polymers comprising POSS, boronic acid, or boronate moieties. As demonstrated by the above minireview, amphiphilic copolymers possessing POSS, boronic acid, or boronate moieties are of high capability to self-assembly in solution systems, because new driving forces such as POSS stacking or B-N dative bond emerge. These hybrid nanoassemblies generally have controllable morphologies, diameters, properties, and shown potential applications in sensor and biomedical areas.

Notably, we should also concern the shortcomings of the nanoassemblies derived from amphiphilic polymers containing POSS, boronic acid, or boronate moieties. (1) POSS cage itself is usually inert and has no evident advantage to endow POSS-based polymers with smart properties; (2) the polymerization degree of POSS monomer is still much lower than normal monomers; (3) boronic acid or boronate moieties incorporated into the main chain of polymers have not been well developed; (4) polymerizable monomers with boronic acid or boronate moieties are rare. Future development directions in this area might focus on the design of novel hybrid polymerizable monomers and explosion of new self-assembly driving forces. For example, new POSS structures with dynamic bond incorporated in the POSS cages may endow POSS-based polymers with smart properties; novel monomers with boronic acid or boronate moieties may endow the polymers with new intermolecular interactions to promote the self-assembly of polymeric networks.

Acknowledgements

The authors like to acknowledge the support of National Natural Science Foundation of China (50873082, 50903066, 51103123, 51273164, U1205113, 51373142, 51403176, 51573150, 51673161); Scientific and Technological Innovation Platform of Fujian Province (2014H2006); Foundation of NSFC-CNRS-PICS (51511130130, PICS 07002 pour le CNRS); National Science and Technology Ministry (2014BAF08B03); The Natural Science Foundation of Fujian Province (2016J01257, 2015J01220); Enterprise Technology Innovation Project of Fujian Province ([2016]411); Key Project of Fujian Department Science and Technology (2013HZ0005-1); The Scientific and Technical Project of Xiamen (3502Z20150047).

Author details

Conghui Yuan[1,2*], Yiting Xu[1,2], Birong Zeng[1,2], Weiang Luo[1,2], Guorong Chen[1,2], Jie Mao[1], Cheng Liu[1] and Lizong Dai[1,2*]

*Address all correspondence to: lzdai@xmu.edu.cn and yuanch@xmu.edu.cn

1 College of Materials, Xiamen University, Xiamen, China

2 Fujian Provincial Key Laboratory of Fire Retardant Materials, Xiamen University, Xiamen, China

References

[1] Dawei Li, Yuguang Niu, Yanyu Yang, Xing Wang, Fei Yang, Hong Shen, Decheng Wu. Synthesis and self-assembly behavior of POSS-embedded hyperbranched polymers. Chem. Commun. 2015;**51**(39):8296-8299.

[2] Kousuke Tsuchiya, Hitoshi Arai, Yoshihito Ishida, Atsushi Kameyama. Dynamic network formation of POSS-pendanted polymer via cage scrambling mediated by fluoride ion. Macromolecules 2015;**48**(6):1636-1643.

[3] Yu-Feng Zhu, Wei Liu, Meng-Yao Zhang, Yu Zhou, Yu-Dong Zhang, Ping-Ping Hou, Yu Pan, Zhihao Shen, Xing-He Fan, Qi-Feng Zhou. POSS-containing jacketed polymer: hybrid inclusion complex with hierarchically ordered structures at sub-10 nm and angstrom length scales. Macromolecules 2015;**48**(8):2358-2366.

[4] Parameswara Rao Chinnama, Hanjun Zhang, Stephanie L. Wunder. Blends of pegylated polyoctahedral silsesquioxanes (POSS-PEG) and methyl cellulose as solid polymer electrolytes for lithium batteries. Electrochimica Acta. 2015;**170**(10):191-201.

[5] Shiao-Wei Kuo, Feng-Chih Chang. POSS related polymer nanocomposites. Progress Polymer Sci. 2011;**36**(12):1649-1696.

[6] Zhenghe Zhang, Yudong Xue, Pengcheng Zhang, Axel H. E. Müller, Weian Zhang. Hollow polymeric capsules from POSS-based block copolymer for photodynamic therapy. Macromolecules 2016;**49**(22):8440-8448.

[7] Wenchao Zhanga, Giovanni Camino B, Rongjie Yang. Polymer/polyhedral oligomeric silsesquioxane (POSS) nanocomposites: an overview of fire retardance. DOI: 10.1016.

[8] Weian Zhanga, Axel H.E. Müller. Architecture, self-assembly and properties of well-defined hybrid polymers based on polyhedral oligomeric silsequioxane (POSS). Progress Polymer Sci. 2013;**38**(8):1121-1162.

[9] Lingnan Chen, Birong Zeng, Jianjie Xie, Shirong Yu, Conghui Yuan, Yinyin Pan, Weiang Luo,, Xinyu Liu, Kaibin He, Yiting Xu, Lizong Dai. A metal-sensitive organic-inorganic hybrid surfactant: POSS-capped dipicolinic acid-functionalized poly(ethylene glycol) amphiphile. Reactive & Functional Polymers. 2013;**73**(8):1022-1029.

[10] Shasha Li, Yong Liu, Sha Ji, Zheng Zhou, Qifang Li. Synthesis and self-assembly behavior of thermoresponsive poly(oligo(ethylene glycol) methyl ether methacrylate)-POSS with tunable lower critical solution temperature. Colloid and Polymer Science. 2014;**292**(11):2993-3001.

[11] Jin Wu, Xiaoyan Song, Lintao Zeng, Jinfeng Xing. Synthesis and assembly of polyhedral oligomeric silsesquioxane end-capped amphiphilic polymer to enhance the fluorescent intensity of tetraphenylethene. Colloid and Polymer Science. 2016;**294**(8):1315-1324.

[12] S. Yusa, S. Ohno, T. Honda, H. Imoto, Y. Nakao, K. Naka, Y. Nakamura, S. Fujii. Synthesis of silsesquioxane-based element-block amphiphiles and their self-assembly in water. Rsc Advances. 2016;**6**(77):73006-73012.

[13] Yuanming Deng, Julien Bernard, Pierre Alcouffe, Jocelyne Galy, Lizong Dai, Jean-Francois Gérard. Nanostructured hybrid polymer networks from in situ self-assembly of RAFT-synthesized POSS-based block copolymers. Journal of Polymer Science, Part A: Polymer Chemistry. 2011;**49**(20):4343-4352.

[14] Cangjie Yang, Yuanming Deng, Birong Zeng, Conghui Yuan, Min Chen, Weiang Luo, Jie Liu, Yiting Xu, Lizong Dai. Hybrid amphiphilic block copolymers containing polyhedral oligomeric silsesquioxane: synthesis, characterization, and self-assembly in solutions. Journal of Polymer Science Part A: Polymer Chemistry. 2012;**50**(20):4300-4310.

[15] Yiting Xu, Min Chen, Jianjie Xie, Cong Li, Cangjie Yang, Yuanming Deng, Conghui Yuan, Feng-Chih Chang, Lizong Dai. Synthesis, characterization and self-assembly of hybrid pH-sensitive block copolymer containing polyhedral oligomeric silsesquioxane (POSS). Reactive & Functional Polymers. 2013;**73**(12):1646-1655.

[16] Yiting Xu, Jianjie Xie, Lingnan Chen, Conghui Yuan, Yinyin Pan, Ling Cheng, Weiang Luo, Birong Zeng, Lizong Dai. A novel hybrid random copolymer poly(MAPOSS-co-NIPAM-co-OEGMA-co-2VP): synthesis, characterization, self-assembly behaviors and multiple responsive properties. Macromolecular Research. 2013;**21**(12):1338-1348.

[17] Yiting Xu, Jianjie Xie, Lingnan Chen, Hui Gao, Conghui Yuan, Cong Li, Weiang Luo, Birong Zeng, Lizong Dai. Synthesis, characterization, and temperature-responsive behaviors of novel hybrid amphiphilic block copolymers containing polyhedral oligomeric silsesquioxane. Polymers for Advanced Technologies. 2014;**25**(6):613-623.

[18] Yiting Xu, Ying Cao, Jianjie Xie, Qi Li, Xianming Chen, Shiao-Wei Kuo, Lizong Dai. Mixed micelles from synergistic self-assembly of hybrid copolymers with charge difference electrostatic interaction induced re-organization of micelles from hybrid copolymers. Journal of Materials Research. 2016;**31**(14):2046-2057.

[19] Birong Zeng, Yueguang Wu, Qilong Kang, Ying Chang, Conghui Yuan, Yiting Xu, Feng-Chih Chang, Lizong Dai. Metal-ions directed self-assembly of hybrid diblock copolymers. Journal of Materials Research. 2014;**29**(22):2694-2706.

[20] Libor Matejka, Miroslav Janata, Josef Plestil, Alexander Zhigunov, Miroslav Slouf. Self-assembly of POSS-containing block copolymers: fixing the hierarchical structure in networks. Polymer. 2014;**55**(1):126-136.

[21] Zhenghe Zhang, Lizhi Hong, Jinxia Li, Feng Liu, Haibo Cai, Yun Gao, Weian Zhang. One-pot synthesis of well-defined amphiphilic alternating copolymer brushes based on POSS and their self-assembly in aqueous solution. Rsc Advances. 2015;**5**(28):21580-21587.

[22] Lizhi Hong, Zhenghe Zhang, Yuan Zhang, Weian Zhang. Synthesis and self-assembly of stimuli-responsive amphiphilic block copolymers based on polyhedral oligomeric silsesquioxane. Journal of Polymer Science Part A: Polymer Chemistry. 2014;**52**(18):2669-2683.

[23] Kun Wei, Lei Li, Sixun Zheng, Ge Wang, Qi Liang. Organic-inorganic random copolymers from methacrylate-terminated poly(ethylene oxide) with 3-methacryloxypropylheptaphenyl polyhedral oligomeric silsesquioxane: synthesis via RAFT polymerization and self-assembly behavior. Soft Matter. 2014;**10**(2):383-394.

[24] Weizhong Yuan, Tianxiang Shen, Xu Liu, Jie Ren. Star-shaped inorganic-organic hybrid polymers with polyhedral oligomeric silsesquioxane core: synthesis, self-assembly and tunable thermoresponse. Materials Letters. 2013;**111**:9-12.

[25] Lei Li, Beibei Lu, Jianning Wu, Qikui Fan, Xuhong Guo, Zhiyong Liu. Synthesis and self-assembly behavior of thermo-responsive star-shaped POSS-(PCL-P(MEO(2)MA-co-PEGMA))(16) inorganic/organic hybrid block copolymers with tunable lower critical solution temperature. New Journal of Chemistry. 2016;**40**(5):4761-4768.

[26] Lei Li, Beibei Lu, Qikui Fan, Jianning Wu, Lulu Wei, Jun Hou, Xuhong Guo, Zhiyong Liu. Synthesis and self-assembly behavior of pH-responsive star-shaped POSS-(PCL-P(DMAEMA-co-PEGMA))(16) inorganic/organic hybrid block copolymer for the controlled intracellular delivery of doxorubicin. Rsc Advances. 2016;**6**(66):61630-61640.

[27] Debashish Roy, Brent S. Sumerlin. Glucose-sensitivity of boronic acid block copolymers at physiological pH. ACS Macro Lett. 2012;**1**:529-532.

[28] Weizhong Yuan, Tianxiang Shen, Jinju Wang, Hui Zou. Formation-dissociation of glucose, pH and redox triply responsive micelles and controlled release of insulin. Polymer Chemistry. 2014;**5**:3968-3971.

[29] Debashish Roy, Jennifer N. Cambre, Brent S. Sumerlin. Triply-responsive boronic acid block copolymers: solution self-assembly induced by changes in temperature, pH, or sugar concentration. Chemistry Communication. 2009:2106-2108.

[30] Kyoung Taek Kim, Jeroen J. L. M. Cornelissen, Roeland J. M. Nolte, Jan C. M. van Hest. A polymersome nanoreactor with controllable permeability induced by stimuli-responsive block copolymers. Advanced Materials. 2009;**21**:2787-2791.

[31] Zhifeng Xu, Khan Mohammad Ahsan Uddin, Lei Ye. Boronic acid terminated thermo-responsive and fluorogenic polymer: controlling polymer architecture for chemical sensing and affinity separation. Macromolecules. 2012;**45**:6464-6470.

[32] Yanet Elised Aguirre-Chagala, José Luis Santos, Bethsy Adriana Aguilar-Castillo, Margarita Herrera-Alonso. Synthesis of copolymers from phenylboronic acid-installed cyclic carbonates. ACS Macro Letters. 2014;**3**:353-358.

[33] Mitsuru Naito, Takehiko Ishii, Akira Matsumoto, Kanjiro Miyata, Yuji Miyahara, Kazunori Kataoka. A phenylboronate-functionalized polyion complex micelle for ATP triggered release of siRNA. Angewandte Chemie International Edition. 2012;**51**:10751-10755.

[34] Fanny Coumes, Patrice Woisel, David Fournier. Facile access to multistimuli-responsive self-assembled block copolymers via a catechol/boronic acid ligation. Macromolecules. 2016;**49**:8925-8932.

[35] Priyadarsi De, Sudershan R. Gondi, Debashish Roy, Brent S. Sumerlin. Boronic acid-terminated polymers: synthesis by RAFT and subsequent supramolecular and dynamic covalent self-assembly. Macromolecules. 2009;**42**:5614-5621.

[36] Yuan Yao, Xuemin Wang, Tianwei Tan, Jing Yang. A facile strategy for polymers to achieve glucose-responsive behavior at neutral pH. Soft Matter. 2011;**7**:7948-7951.

[37] Erin Sheepwash, Vincent Krampl, Rosario Scopelliti, Olha Sereda, Antonia Neels, Kay Severin. Molecular networks based on dative boron–nitrogen bonds. Angewandte Chemie International Edition. 2011;**50**:3034-3037.

[38] Erin Sheepwash, Nicolas Luisier, Martin R. Krause, Stefanie Noé, Stefan Kubikb, Kay Severin. Supramolecular polymers based on dative boron–nitrogen bonds. Chemical Communication. 2012;**48**:7808-7810.

[39] Longyu Li, Conghui Yuan, Lizong Dai, S. Thayumanavan. Thermoresponsive polymeric nanoparticles: nucleation from cooperative polymerization driven by dative bonds. Macromolecules. 2014;**47**:5869-5876.

[40] Conghui Yuan, Ying Chang, Jie Mao, Shirong Yu, Weiang Luo, Yiting Xu, S. Thayumanavan, Lizong Dai. Supramolecular assembly of crosslinkable monomers for degradable and fluorescent polymer nanoparticles. Journal of Materials Chemistry B. 2015;**3**:2858-2866.

Chapter 6

Introduction to Electronic Properties and Dynamics of Organic Complexes as Self-Assembled Monolayers

Maddalena Pedio and Barbara Ressel

Additional information is available at the end of the chapter

http://dx.doi.org/10.5772/68111

Abstract

Self-assembled monolayers (SAMs) of organic-conjugated transition metal complexes on surfaces is a focus of both device engineering and basic science, since it is a key factor in nearly all important aspects of device performances, including operation voltages, degradation, and efficiency. The huge amount of literature results related to the first monolayer, and reorganization and self-assembling processes are due to the general accepted result that structural and chemical properties of the first monolayer are the key parameters for controlled thin film growth. Optical and magneto-electronic properties are intimately connected, and the accurate determination of electronic levels, excitation, and relaxation dynamics is mandatory for the optimization of electronic, photovoltaic, and opto-electronic devices. Quite a number of electronic states is generated by the interaction of light with complex organic molecules. Time-resolved spectroscopies are a new investigation tool that gives the possibility of correctly addressing their origin and life time. Examples of prototypical systems are presented and discussed. We review on complementary techniques, trying to single out how different approaches are fundamental to fully characterize these complex systems.

Keywords: self-assembled monolayer (SAM), surface structures molecular layers, nanotechnology, electronic properties, spectroscopies, time resolved

1. Introduction

The fundamental understanding of the molecular solids properties and the controlled ordering processes of organic complexes received a paramount interest for their promising applications. The current challenge in nanostructured materials consists of the understanding of ever smaller scale and in exploiting their functionality at the relevant length, time, and energy scales.

In organic semiconductors, their molecular aggregation and crystal packing in solid state induce unique properties that differ from those of the single molecule components or their liquid solutions, affecting their structural, optical, and electronic-magnetic properties. The understanding and the control of the processes determining the dynamic properties of organic semiconductors are crucial for further development of organic electronics and other technological applications.

The level of effort placed on the study of transition metal complexes (TMCs) [1] and their single layers on surfaces has grown steadily in the last few decades. Large molecules deposition on surfaces leads to a high variety of interactions. The bond strength spreads from very weak VdW to no covalent or strong chemical bonding. Normally, organic-inorganic are multiphase systems with complex phase diagrams. The experience in these decades has stated that molecule-substrate must be viewed as a whole. Even in weak-bounded molecular layers [2], the simple change of symmetry induced by adsorption and the induced dipoles can lead to modifications of the molecular and metallic states, affecting the electronic structure at the interface [3–6]. The static and dynamic characterization on different time scales are crucial for the comprehension of such complex systems, as well as the understanding of the mutual correlation among the structural, opto-magnetic, and electronic properties. In view of applications, thin and ultrathin films of TMCs are of great current interest, and the control of the thin films' growth and crystallinity is of crucial importance to optimize the performance of future electronic devices. To acquire the ability to design materials with selected properties requires a deep understanding of the proper molecular precursors characteristics used as building blocks and of the modifications induced when deposited in films.

In this chapter, case systems involving TMCs will be addressed, focusing the discussion on their dynamical properties.

Among TMCs, the tetrapyrrole complexes, such as metallo-porphyrins, phthalocyanines, and dipyrrins, are ubiquitous molecules, and their presence spans the geological world with etioporphyrins and up to biology with hemoproteins. They serve as agents to transport molecules (hemoproteins) or electrons (chlorophylls) and are also important as photochemical agents with chlorophyll antennae. Synthetic porphyrins and phthalocyanines are used in the phototherapy of cancer or in solar cells, through virtue of their unique photophysical properties, and are versatile building blocks for the realization of molecular materials, which exhibit various light-driven chemical reactions. A multitechnique approach is necessary to extract and enlighten the interrelation between structural details and the opto-electronic properties [7]. A number of reviews [3–6, 8, 9] and accounts dealing with specific aspects of organic layers on different classes of substrates, conductive or semiconductors are present in the literature.

Among the organic semiconductors, those formed by tetrapyrrole metal complexes (such as porphyrins and phthalocyanines) are found in a wide variety of technological applications including electronics, biological, and chemical contexts. Porphyrins often act as metal binders and find applications in dye-sensitized solar cells, in molecular magnets, oxygen sensors, etc. To fabricate organic photovoltaic (OPV) solar cells [10] with high conversion efficiency, it has been proved that it is crucial to control the crystal structure, crystallinity, and molecular

http://dx.doi.org/10.5772/68111

orientation of the thin films used, because the diffusion length of excitons and/or charge carriers, electrons, and holes in their crystalline films becomes longer than in their amorphous form. These factors became even more crucial in the single layer interacting with substrates, as discussed in the next paragraph.

We review the recent findings on the excitation and relaxation dynamics of TM tetrapyrrole single layer (metal octaethyl porphyrins, MOEP; metal tetraphenyl porphyrins, MTPP) and phthalocyanines (MPcs) by means of time-resolved optical and photoelectron spectroscopies, enlightening the aspects of molecular reorganization chemistry on model solid surfaces, with a focus on molecule-substrate bond and energy level alignment.

2. From single molecules to aggregates

The knowledge of the molecular properties in the gas phase is mandatory for the understanding of the whole set of molecular applications. The spectroscopic characterization of molecular energy levels in vapor has had enormous improvements in the last decades [11]. In TMC, the central metal ion and its n*d* orbital occupancy strongly affect the complex properties. For example, in transition metal phthalocyanine (MPc) [12], the comparison between the series of TMCs versus 3d occupancy shows that in choosing the proper metal, the filled orbitals and in particular the highest occupied molecular orbital (HOMO) allow to tune their binding energy to engineer their transport and optical gaps.

In organic semiconductors, the solid aggregation induces unique optical and electronic-magnetic properties, which often differ from those of the single molecule components or their liquid solutions, affecting their structural, optical, and electronic-magnetic properties as discussed in the following sections. The reason is qualitatively illustrated in **Figure 1**, where the energies of the so-called frontier orbitals, highest occupied molecular orbital HOMO and the lowest unoccupied molecular orbital (LUMO) result perturbed in solids due to intermolecular interactions.

Strong theoretical efforts have been applied in the last decades to characterize the geometry, electronic, and optical properties of the TMC, accompanied by a robust experimental set of

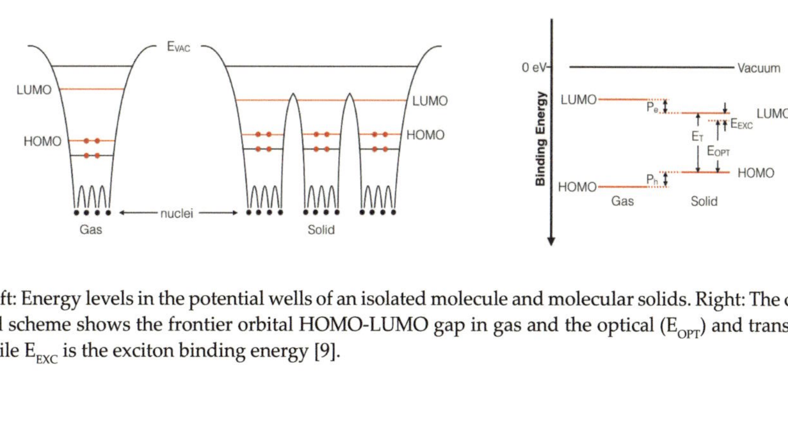

Figure 1. Left: Energy levels in the potential wells of an isolated molecule and molecular solids. Right: The corresponding energy level scheme shows the frontier orbital HOMO-LUMO gap in gas and the optical (E_{OPT}) and transport (E_T) gaps in solid, while E_{EXC} is the exciton binding energy [9].

data on free molecules, solutions, and solid aggregates [4, 13, 14]. The aggregation enhances the intermolecular π-π stacking, facilitating the intramolecular charge transport. The understanding and the control of the processes determining the magneto-electronic and dynamic properties of organic semiconductors are crucial to the further development of organic electronics and other technological applications. Better crystal quality with large domains increases properties like conductivity by more than one order of magnitude. It has been proved how critical it is to control the crystal structure, crystallinity, and molecular orientation of the thin films because the diffusion length of excitons and/or charge carriers, electrons, and holes in their crystalline films becomes longer than in their amorphous form. Intermolecular interactions, dishomogeneities, and grains within the molecular solids are related to the π-π intermolecular interactions exhibiting strong electron-phonon coupling, important for the charge and exciton transport in organic devices [15].

Molecular crystalline materials can be altered by manipulating spin, electron transfer, proton transfer, molecular structure, and orientation, leading to dynamical switchable properties [16] by external stimuli such as light, electric field, and temperature. The precise control of the switching properties is the focus of recent frontier developments. For single layer, the dynamics study is still to be systematically understood due to the experimental difficulties. Recent results will be discussed in Section 4.

2.1. Transition metal complexes

Transition metal coordination complexes are conjugated molecules that consist of a central metal atom or ion (coordination center) and a surrounding array of bound molecules or ions, ligands, or complex agents. Common coordination geometries are shown in **Figure 2**. Properties of TMCs include absorption and transmission of visible light (highly colored), multiple oxidation states of the metals, paramagnetism dependent on the *nd* electronic configuration of the metal ion, metal oxidation state and on ligand field, reactivity.

Here, we mainly focus on the large planar macrocycle complexes, tetrapyrrole molecules metallo porphyrins (MOEP, MTPP), and phthalocyanines (MPcs), whose specific functionalities are related to the presence in the macrocycle of the metal center in **Figure 2**. Porphyrin derivative complexes are ubiquitous molecules in nature and present an enormous variety of properties [10, 17]. This pervasiveness arises from their special tetrapyrrole planar structure

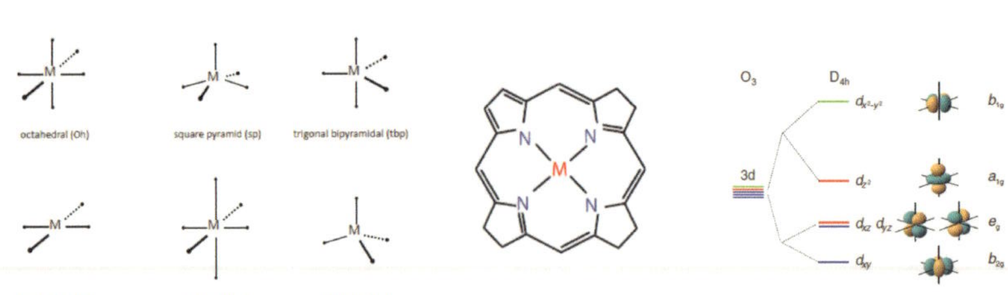

Figure 2. Left: Common coordination geometries in TM complexes and the structure of the porphyrin macrocycle common to the tetrapyrrole complexes. Right: The 3d orbital splitting induced by a D4h square planar crystal field, adapted with permission from Ref. [18], copyright 2012 American Physical Society.

http://dx.doi.org/10.5772/68111

with a metal atom bound to four N atoms in the macrocycle and from its stable π-conjugated system. The organic ligand of metallo-organic molecules affects the electronic states of the metallic ion. Central metallic ions atomic orbitals are hybridized with the organic ligand in the macrocycle, leading to molecular orbitals with a partial metallic character. The 3d orbitals of the transition metal ion get split by the molecular tetragonal D4h symmetry crystal field, giving rise to the *3d* character molecular orbitals indicated in **Figure 2** see for example Refs. [18, 19].

The large aromatic nature of tetrapyrroles provides the basis for their unique absorption properties, for example, the typical UV/Vis spectra of porphyrins exhibit absorptions in two regions termed the Soret or B band (~380–420 nm), corresponding to transition from ground state to the second excited state, and the Q bands (~500–800 nm) related to the first excited state (**Figure 3**). This is a result of the splitting of the main frontier molecular orbitals, the highest occupied molecular orbital (HOMO) and the lowest unoccupied molecular orbital (LUMO). The electronic structure of porphyrins can be described by the semi-quantitative Gouterman's *four-orbital model* that considers the two HOMOs (labelled as a1*u* and a2*u*) and LUMOs (labelled *eg*) of the porphyrin (Hückel theory) and mixes the four possible optical excitations between them using configuration interaction theory to account for electron interaction [17].

The surrounding groups present around the porphyrin ring have a detectable effect on the binding energy and shape of the molecular orbitals and the *3d* electronic configuration through conjugation. This is reflected in spectroscopic data as discussed, for instance, in the case of Ni complexes [21], where core level photoemission and X-ray absorption spectra of Ni, C, and N indicate that different charge transferred occurs in the different tetrapyrrole complexes, at the metal-to-ligand bonds, from the occupied Ni 3 d_{xz}, d_{yz} (e_g) orbitals to the unoccupied N $2\pi\pi^*$ (e_g) orbitals, hence inducing changes in the back donation among the different molecules, from NiOEP to NiPc. Other examples are present in the literature, for example, for Fe complexes [22].

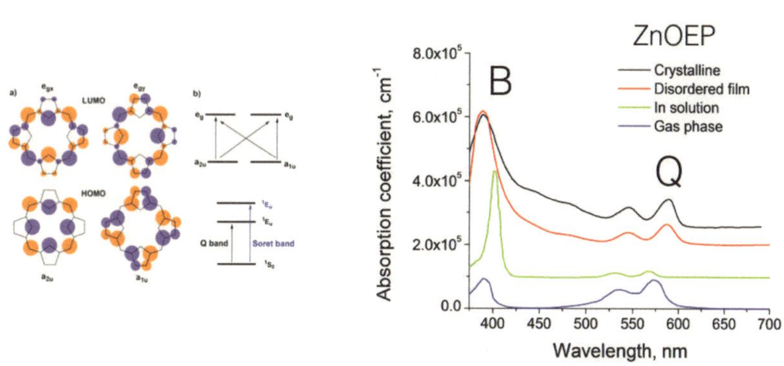

Figure 3. Simplified representation of the one electron transitions from HOMO to LUMO, related to their absorption spectra B (Soret) and Q band regions (from ref [17]) in the *four orbital-model*. Example of crystalline and disordered optical absorption for ZnOEP compare with the ZnOEP in toluene solution and gas phase data [20].

Tetrapyrrole derivatives aggregate [23] into a plethora of nanostructures such as films, crystals, tubes, rods, wires, nanoparticle spheres, and complex fractal-like or chiral patterns. By varying the choice of the metal center, the bonding and characteristics of the metalloporphyrin may be dramatically affected. Systematic experimental and theoretical evaluation of the porphyrin and MPc enlighten the role of the 3*d* metal electronic states and occupancy in the conjugation of these complexes [18, 19, 24–27].

2.2. Organic thin films

The structural and electro-optical properties present a strong mutual influence and fine details' variation that can lead to strong alteration of macroscopic properties in the aggregates [28] as the transport and optical gap or the magnetic moments in the paramagnetic complexes. Moreover, organic thin films present structural polymorphism [29], depending on the molecular structure and the strength of the intermolecular interaction. Several works demonstrate the effect of different stacking structures at macroscopic scales [19, 30]. Octaethyl porphyrins (OEP) for example present a simple structure with the eight external ethyl groups in alternate orientation. Even in this relatively simple structure, the ruffling of the ethyl groups in aggregation or self-organization induces perturbation on the electronic properties as shown theoretically [31] for NiOEP. Moreover, the ordering is a crucial factor for applications: differences in optical absorption have been measured when comparing, for example, ZnOEP films with different degrees of long range order [32].

2.3. Self-organization in the first layer

The interest in the first monolayer of (organic π-conjugated systems or fullerene) molecules on single crystals adsorption is due to the general accepted result that structural and chemical properties of the first monolayer are keys for controlled film growth. Adsorption of complex molecules on metal and semiconductor interfaces may lead to the formation of highly ordered nanostructuration that can then be exploited for the fabrication of surface-supported nanostructures as molecular building blocks confined to 2D systems. Self-assembled monolayers are ordered molecular assemblies formed by the adsorption onto an inorganic or organic (heterojunctions) substrate. Self-assembly [33], in a general sense, might be defined as the spontaneous formation of complex hierarchical structures from pre-designed building blocks, typically involving multiple energy scales and multiple degrees of freedom. The knowledge of the interaction details of the first organic layer is mandatory: the investigation on the perturbation induced by the molecule-substrate interaction and how the isolated molecule properties are perturbed by intermolecular interactions in the assembling. Absorption and electron spectroscopies, together with scanning tunneling microscopy (STM), are precious tools that provide a clear framework of these complex systems.

In these studies, the question to be answered concerns the types of structures and phases that are formed and which parameters characterize the order:

- the self-assembling driving forces (what determines the growth kinetics and the growth regimes),

- what are the *internal* (e.g., molecular conformers variation or substrate orientation) and the *external* (e.g., temperature) control parameters?
- the nature of the phase transitions and how do the various degrees of freedom and the different constituents of the molecule (headgroup, chain or backbone, endgroup) have an impact on the growth and the structure.

3. Mutual effects of self-assembling on substrates and molecular conformers in tetrapyrrole complexes

Organic films can be deposited by a variety of methods, as spin coating from solutions or ultrahigh vacuum (UHV) deposition. Here, we refer mainly to the latter method. Porphyrins and phthalocyanines, due to their very low room temperature and vapor pressure (10^{-14} Torr), can be sublimated in UHV and thus prepared and characterized as controlled grown molecular layers.

In the self-assembled monolayer (SAM), on ordered substrates, the break of translational symmetry at the interface induces a variation of the boundary conditions with relevant variation of the charge densities both in the molecule and in the substrate. The wave functions of these interface states have a large amplitude at the interface and decay quickly into both materials [3]. The mechanism of molecule surface interaction can be quite complex due to the balance between intermolecular binding forces and molecule-substrate interactions, involving a large number of sites due to the molecular size with respect to the unit cell of the substrates. This can result in either displacive substrate reconstructions and/or in complex structural deformation of the adsorbed molecules. The molecule-substrate system must be taken into account as a whole, and it is not possible to simply transfer molecular functionalities. The simulation of such systems is still a challenging task and deserves state-of-the art theoretical framework.

The tendency to form ordered self-organized structures on a variety of substrates and the existence of extended π-electron systems makes the study of tetrapyrrole TM complexes important from both fundamental and technological viewpoints.

In his recent detailed review, Gottfried [13] stressed that by means of the conjugated organic molecules, adsorption on metal substrates is possible to engineer the modification of interfacial properties, especially for planar tetrapyrrole complexes. The understanding of the interface electronic structure is therefore a prerequisite for insightful materials design.

3.1. Ordering in the first layer and changes in molecules

The self-assembly of TM complexes is driven by intermolecular interactions, which can be attractive or repulsive, depending on the molecular structure and the character of the surface chemical bond ([4, 5, 13] and references therein). An interface dipole takes place even in weakly interactive SAM interfaces (see also **Figure 5**), while chemical bonds facilitate the electron transfer between substrate and adsorbate. The resulting dipoles along the surface normal are repulsive and can over-compensate the lateral Van der Waals attraction. The preferred orientation is with the macrocycle parallel to the surface on conductive substrates. The lateral interactions are closely coupled to the vertical adsorbate-substrate distances, though their leading determination implies important experimental efforts. Other important aspects include the degree of mismatch between the molecular and the substrate symmetries, which can lead to adsorption-induced symmetry reductions in the molecule, and substrate-induced template effects, which affect the supramolecular arrangement. Moreover, in case of peripheral non-planar groups, the adsorption leads to a distortion of the molecular conformers as in MTPP, either because of their inherent non-planarity or because of intramolecular sterical hindrance. In the adsorbed state, this leads to a competition between the energy-decreasing maximization of the adsorbate-substrate contact, which forces the molecule to approach a planar conformation, and the energy-increasing intramolecular strain resulting from the distortion toward planarity.

As in isolated molecules, the central atom's electronic configuration and 3*d* occupancy play a major role in the structural rearrangements of selected SAM [34] and in molecule-substrate bonds (chemisorbed/physisorbed) as found in MPc and porphyrins (review Table 4 Ref. [13]). Case systems are discussed in Section 3.3.

3.2. Electronic structure and density of state at SAM interfaces

The characterization of SAM by absorption and electron spectroscopies, together with scanning microscopies, provides unique information on the reorganization of the density of electron states (DOSs) at interfaces ([35, 36] and refs therein). At interface, several processes take place in the delicate intermolecular, molecule-substrate interaction balance [37]: correlations between strength of molecule/metal interaction, average bonding distances, adsorption-induced molecular conformation changes leading to intramolecular dipoles, organic/metal interface dipoles, and the resulting unexpected energy level alignment. Interface energy level alignment, band bending, chemical reaction, charge transfer (CT), metal-organic interface dipole formation, internal molecular dipole, and Debye screening are all necessary factors to describe the single layer on surfaces. The reader is addressed to the numerous excellent papers and reviews present in the literature to get more details, for example, Refs. [4–6].

For the aim of this review, we just need to know that PES is a surface-sensitive technique that provides a direct insight of occupied density of states of electronic energy levels. It relies on the photoelectric effect: when a photon, whose energy $h\nu$ is larger than the sample work function (WF or ionization potential in case of gases), is shone on a sample, an electron is ejected with kinetic energy $KE = h\nu - EB - WF$, where EB is the electron's binding energy. The combination of photoemission with inverse photoemission allows one to measure the electron affinity (EA), the work function (WF), the transport gap, (E_T), and the interface dipole (Δ), as indicated in **Figure 4** [6]. In organic semiconductors, the ground state is characterized by the presence of electrons in the highest occupied molecular orbital (HOMO) and vacancies in the lowest unoccupied molecular orbital (LUMO). The absorption of a visible photon creates an

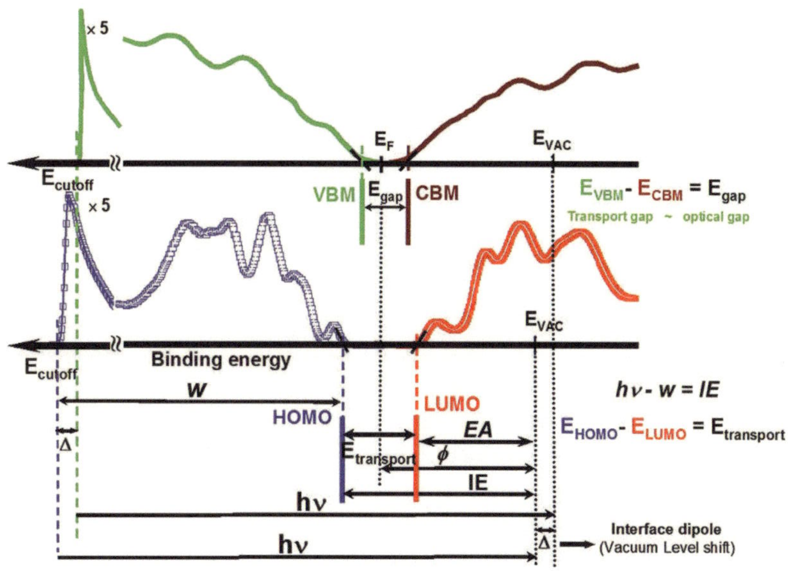

Figure 4. Combined photoemission and inverse photoemission spectra of an inorganic solid (upper panel) and thin organic film (lower panel) used to define the relevant quantities. Reprinted with permission from Ref. [6]. Copyright 2007–2017, American Chemical Society.

excited state, where an electron is promoted in the LUMO and a vacancy is left in the HOMO. However, this state is not stable because its energy is higher than that of the ground state and therefore it will decay.

Among the techniques that give access to magnetic moments, the X-ray magnetic dichroic spectroscopy measurements with elemental sensitivity is widely used in the case of paramagnetic SAM [18, 19, 38].

Moreover, imaging scanning microscopies such as scanning tunneling microscopy (STM) and scanning tunneling spectroscopy (STS) are giving a huge contribution to reveal the details of organic-inorganic systems. Among the excellent reviews present in the literature, we cite Ref. [39] dedicated to TMC interfaces.

Finally, a specific field of investigation is reserved to interface build by paramagnetic complexes, the spinterfaces, that deserves experimental methods sensible to the magnetic properties of the system. The study of the interconnections between the electronic configuration of the central atoms in the adsorbed TMC, the details of the interaction, and the resulting magnetic properties has recently produced new important results, useful in magneto-optical devices.

3.3. Case systems and trends

The tetraphenylporphyrin (TPP) complexes deposited on metal surfaces are one of the most studied self-organized systems in the literature. They present the tendency to polymorphism related to the external phenyl orientations and the consequent perturbation of the macrocycle planarity. A general trend can enlighten the interaction of the adsorbed TPP: weak interaction on Ag and Au; the intermolecular interactions result in attractive intermolecular forces, leading to the formation of close-packed islands with structures mainly following the molecular symmetry, while the adsorption on more strongly interacting substrates (Cu(110)), though still related to attractive intermolecular forces, leads to more complex structures. In case of metal-free TPPs on strongly interacting substrates, intermolecular repulsion takes place. This trend is common to other porphyrins.

Confirmation of complex interaction frameworks at SAM interfaces is found in the weakly interacting porphyrinoid layers deposited on conductive substrates, such as Au(111) and graphite, HOPG [2], where dispersion is found to play a critical role in altering the adsorption and charge distribution. The comparative study between the two different substrates demonstrates that CoOEP bind more strongly to gold than to graphite, enlightening the molecule-substrate interactions in weakly bond interfaces and the role of dispersion forces between molecules and conductive substrates. Charge redistribution maps of both interfaces display charge localization mostly on the porphyrin molecule indicating a "push back effect" from the substrate, that is, a depletion of charge in the topmost layers. Density functional theory (DFT) calculations with dispersion indicate a larger substrate-to-molecule charge push on Au(111) than on HOPG (**Figure 5**). Moreover, comparison of density of states of isolated CoOEP molecule and its monolayer on gold-HOPG substrates revealed significant orbital hybridization and band shifts. Dispersion interactions played a critical role in altering the adsorption and charge distribution, especially in the Au(111) system. This framework could be valid also in other weakly bond organic-inorganic systems and should have important effects on the dynamical processes.

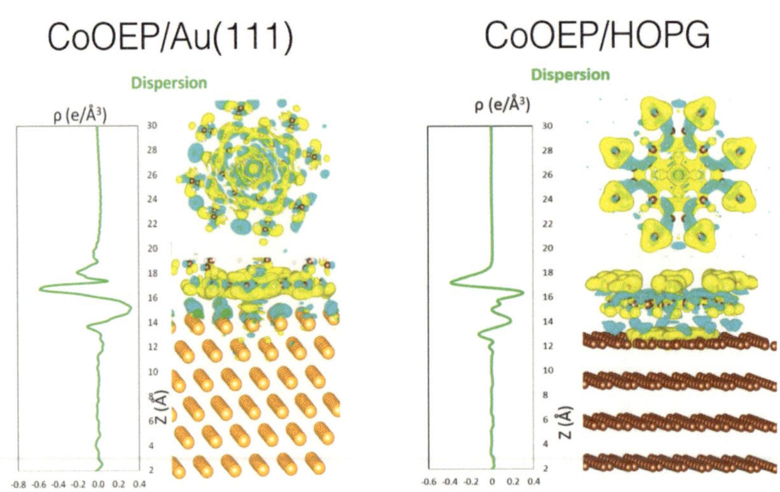

Figure 5. The charge density difference for CoOEP/Au(111) (left) and CoOEP/HOPG (right) systems. For each system, the panel on the right represents the 3D iso-density (+: yellow and −: cyan) of charge density difference. It is reproduced with permission from Ref. [2]. Copyright 2014, Royal Society of Chemistry.

MPcs also show a trend. The tetrapyrrole SAM can self-organize into periodic ordered arrays on ordered surfaces, forming a variety of molecular 1D or 2D superstructures (see Tables 2 and 3 in Ref. [13]) and allowing for tunable anchoring sites. The MPc assembled in compact chains along the Au(110) reconstructed channels representing a dense single layer of molecules with 2D long range geometry where the molecules are all lying on the gold substrate and with little dispersion of the molecule-substrate distance. The interaction of MPc on the Au(110) surface is mediated by the 3*d* metal states occupancy [39]: while Ni, Cu, and Zn metallic centers do not directly interact with the underlying Au substrate, transition-metal center atoms such as Fe and Co induce interaction states close to the Fermi level. The interaction of the regularly spaced Fe and Co metallic centers with a non-magnetic substrate, highly anisotropic Au(110), can break the symmetry of the metallic *d* orbital carrying the magnetic moment, giving rise to reduced magnetic moments (**Figure 6**). In FePc and CoPc, a charge transfer between the metal substrate and the molecular layer induces a redistribution of the molecular orbitals and structural reorganization of the anisotropic Au(110) substrate. In the case of FePc/Au(110) system, it has been systematically shown how the interaction with the surface influences the ordering of the MPc states leading to novel interface properties. These results are intrinsically connected with the intermolecular and molecule-substrate interactions balance produced at interface.

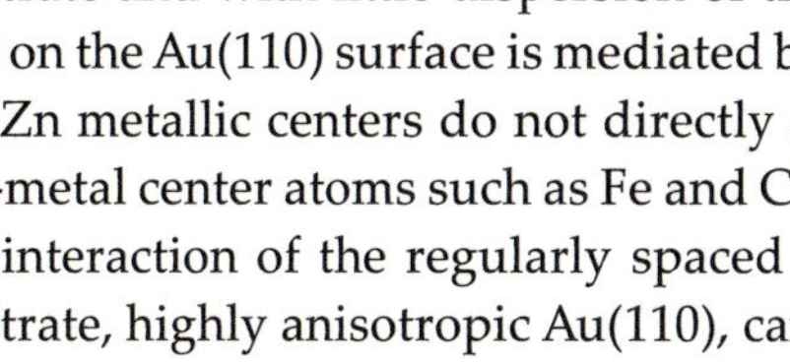

	Easy direction	$\mu_L^{easy}/\mu_L^{hard}$	$\mu_S^{eff\,easy}/\mu_S^{eff\,hard}$
FePc TF	*x*,*y*	2.5	21
CoPc TF	*x*,*y*	5.3	>30
CuPc TF	*z*	4.6	14
FePc SL	*x*,*y*	2.6	>30
CoPc SL			
CuPc SL	*z*	7.5	>30

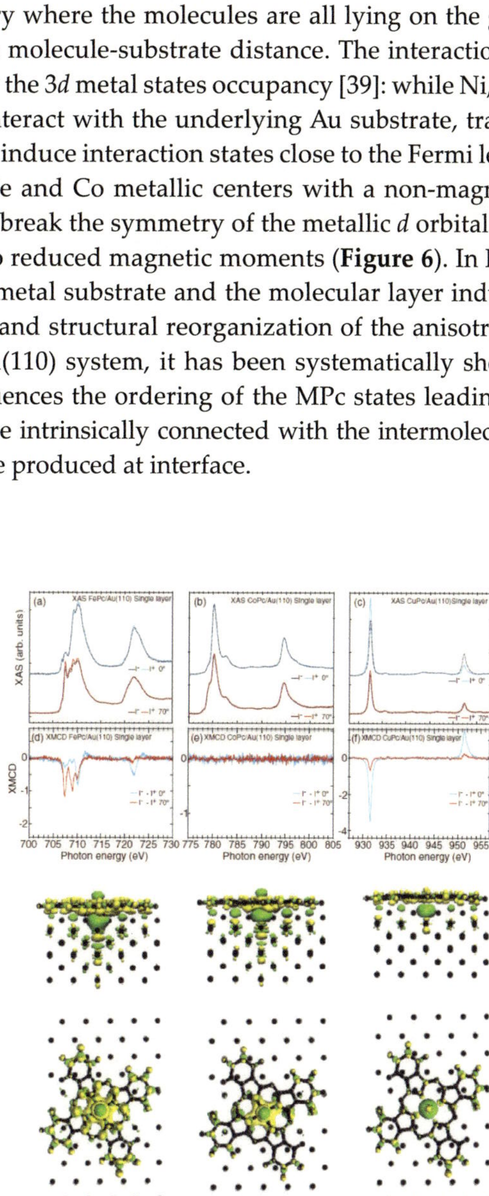

Figure 6. Circularly polarized X-ray absorption spectroscopy metal $L_{2,3}$ spectra (a), (b), (c) and XMCD signals (d), (e), (f) for FePc, CoPc, and CuPc single layers, respectively, collected at 0 and 70° of impinging photons and magnetic field (5 T) at $T = 8$ K. The underneath panels report positive (green) and negative (yellow) charge density transfer upon adsorption of the three MPcs. The table reports the magnetic moments ratio between the easy and the hard magnetization axis at $B = 5$ T, as measured by XMCD. Adapted with permission from Ref. [18], copyright 2012, American Physical Society.

4. Dynamics

In the following paragraph, we will focus on a few selected examples to illustrate the power of the time-resolved investigation methods, when applied to the study of the electronic properties of gases or solid compounds.

For all the already-mentioned reasons, metallo-porphyrins and phthalocyanines are quite interesting systems. To engineer their electronic levels according to different applications either in photovoltaic devices or in opto-electronics (to mention some examples), it is fundamental to study in detail the mechanism of light absorption.

There are different de-excitation channels:

- a radiative decay, where the electron loses energy via the emission of photons of different wavelengths (fluorescence) and
- the creation of an "exciton", an electron-hole bound couple, whose energy is located inside the HOMO-LUMO band gap.

Excitons are crucial for transport in the organic compounds (Section 2.2). If the exciton binding energies are too high, about 1 eV, the bonding lengths are of the order of some Angstrom, so the exciton can move through the system before the recombination of positive and negative charges and thus get lost. But if the exciton is dissociated, that is the bond is broken, the two charges are free and can be collected at the electrodes of a photovoltaic device, for example.

So, several physical steps need to be characterized to engineer a proper molecule with the desired chemical and physical properties to be applied in an effective way to a working device, and time-resolved spectroscopies offer a new insight into all these processes.

Moreover, in order to apply molecular solids in new and fancy devices, based on a reversible chemical or structural modification induced by an external stimulus, like a flash of light, an electric field, or temperature or pressure variations, it is fundamental to understand how these mechanisms are activated. This family of compounds, named "dynamic molecular solids", is discussed in Ref. [16], where the author shows how fundamental is the control of electron, proton, and molecular motion. For example, a photo-induced molecular structural change could be used to control magnetic properties of the molecular solid.

4.1. Dynamic time scales in solids

What happens when a photon is absorbed? The first event after radiation absorption is an increase in energy of the system. In molecular aggregates after excitation, the molecule absorbing light has to redistribute the gained energy. It can be redistributed via different intramolecular processes, that is different channels that involve both atoms (nuclei vibrations) and electrons. All the possible processes have been deeply investigated at the beginning of 1900 by A. Jablonski. Depending on the amount of energy, the molecule can dissociate or isomerize (same atomic configuration but different geometry); otherwise, the energy is redistributed via different processes (see **Figure 7** left):

- internal conversion (IC), which is the transition between electronic states of similar spin,
- intersystem crossing (ISC), which involves states of different spin multiplicities, and
- intramolecular vibration redistribution (IVR), that is vibrational energy flow from a given vibrational mode (or modes) to others.

All these processes are induced by modifications in the electronic structure, like the occupancy of the orbitals or the oxidation state. As can be seen also in **Figure 7** (right), being related to electrons, all these phenomena take place on an ultrafast time scale.

In **Figure 7** right are sketched several electronic/atomic processes and the time scales at which they manifest. Processes involving atoms' nuclei are typically on the pico/nano second time scale, while the time scale of electronic processes is in the femto/atto second region.

To access information on the electronic time scales, it is necessary to use investigation tools with comparable time duration [40] like ultrashort light pulses (of the order of a few tens of femtosecond) and proper wavelength, from the far infrared to the soft X-ray spectral region. With soft X-rays, it is possible to access the energy redistribution processes having the elemental selectivity for both the electronic structure and the geometric structure: the use of the ultrafast techniques to study MPcs and porphyrins has led, in the past decades, to an increased understanding of their photophysical and photochemical properties.

4.2. Principles of selected time-resolved spectroscopies

Since more than 100 years, researchers spent time in finding an investigation technique that was able to provide information on the time scale. In 1878, Leland Stanford, founder of the famous Stanford University, asked the photographer Eadweard Muybridge to demonstrate that while a horse is trotting or galloping, there is a moment in which the four legs are not touching the ground [41]. To prove this statement, the photographer recorded the first slow motion movie of history. He used 12 photocameras in raw that shot an image each, triggered

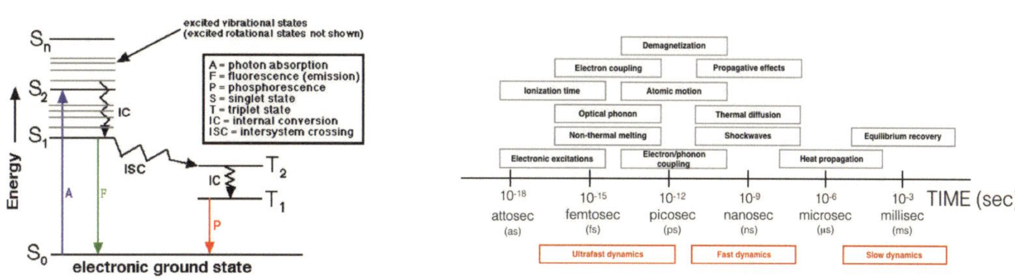

Figure 7. Jablonski diagram, from http://www.shsu.edu/chm_tgc/chemilumdir/JABLONSKI.html and dynamic processes time scales.

by a wire that was pushed down by the wheel of the sulky, carrying the jockey, while it was passing in front of them.

From that time, big improvements have been made on the experimental setup to study the time evolution of a specific phenomenon. The limit of the high-speed photography reached few milliseconds (10^{-3} s) in 1950, but only with the advent of ultrashort laser light pulses at the end of 1990, it was possible to set up a new class of experiments with time resolution better than 10^{-9} s (nanoseconds). The limit of the 10^{-18} s (attoseconds) time scale was reached for the first time in 2001 by Krausz & Co. [42]. With this kind of time resolution, it is now possible to observe and follow in real time chemical reactions on the electrons' time scales.

Going back to the starting example of the horse, we can say that a single picture of the horse is providing the *structural* characteristics of our sample or reaction or any other phenomenon, while the slow motion movie is providing its *kinetics*.

Several spectroscopic techniques are based on the following mechanism: a light pulse is used to excite the sample (pump), while a second light pulse is used to investigate what happened (probe). To realize the *slow motion movie,* we need to control the delay between the pump and the probe pulse, and this is done by varying the spatial distance the two beams have to travel. Since the probe has to reach the sample after the pump has excited it, the optical path of the probe has to be longer. Commonly, a computer-controlled *delay line* varies the path of the pump: when the two optical paths have exactly the same length and the two beams are perfectly overlapping, the light pulses are in temporal and spatial coincidence. This approach can be applied to several materials' characterization techniques. For practical reasons, it started with optical methods, since visible or infrared light pulses are quite easy to achieve with lasers. We are interested in soft X-ray-based techniques like photoelectron spectroscopy (PES). For the description of the technique, see Refs. [6, 37]. In the past 20 years, XUV light pulses became available in ordinary laboratories, thanks to ultrafast table top lasers, exploiting the high-order harmonic generation (HHG) process. For a description of the technique and of the experimental setup, see Refs. [43, 44] and references therein.

In a time-resolved photoemission (TR-PES) experiment, **Figure 8**, a pump pulse is used to excite electrons from filled to empty states, typically by means of an NIR/VIS or UV pulse. Subsequently, an XUV probe pulse is used to emit electrons from the ground state and from the excited state. The kinetic energy of the emitted electrons is then measured varying the delay between the pump and XUV probe pulses [45]. So far, such measurements have been mainly carried out with both the pump and probe energies from the near infrared to UV regimes (up to ~6 eV). In this case, only electrons, which have interacted with photons from both pulses, will be emitted from the sample, and the technique is then often referred to as (time-resolved) two-photon photoemission spectroscopy (2PPE).

4.3. Discussion on selected systems: gas phase dynamics in molecules, thin films, interfaces SAM

The properties and thus the functions of the transition metal complexes (TMCs), even within a single molecular structure, depend specifically on the metal nested at their center. The

http://dx.doi.org/10.5772/68111

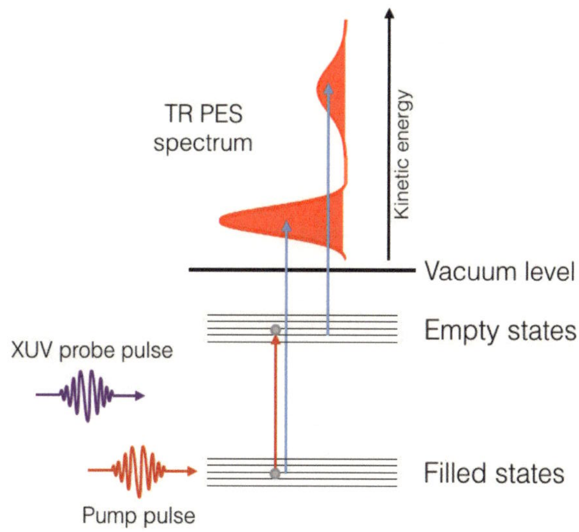

Figure 8. A sketch of the TR-PES process: the pump pulse is used to excite electrons from filled to empty states. Delayed in time, the XUV probe is used to emit electrons from the ground state and from the excited state. The kinetic energy of the emitted electrons is then measured. The measurement is then repeated several times varying the delay between the pump and XUV probe pulse, until the full dynamics of the process is mapped.

energetics and dynamics of the excited states differ among the various metal-porphyrins (characterized by different peripheral groups) and MPc and on the surrounding environment (the presence of a solvent or morphology). Quite critical is the dependence on the atomic number of the metallic atom and on whether the molecular complex is in solution or forms a thin film. Here some examples are discussed.

Gas phase: While the majority of the time-resolved measurements on metal complexes have been performed in solutions, only few literature works are related to the relaxation dynamics of molecules in gas phase. For tetrapyrrole molecules, early works have been published by Gouterman while recently the ultrafast dynamics of porphyrins in gas phase has been performed.

The relaxation dynamics of CuTPP, CuOEP, and H_2TPP were investigated in pump-and-probe experiments [46] based on the optical absorption spectra (seen in **Figure 2**) of Cu porphyrins. The molecules were excited in the Soret B band using 3 eV photon energy (400 nm), while the probe was a 4.5 eV photon (266 nm). The energy diagram of the experiment is shown in **Figure 8** bottom.

Upon excitation, the molecules relax showing different decay times, following the scheme of the picture: a first rapid decay ($t < 100$ fs) from S_2 to the charge transfer (CT) state is followed by another rapid decay in the range of about 300 fs -1 ps to the triplet state, and a final long relaxation to the ground state takes place in the ns time scale. From the spectra of **Figure 9** (top), it can be seen that the experimental data are fit with the convolution of three exponential curves to account for the three different decay times of the process. This behavior

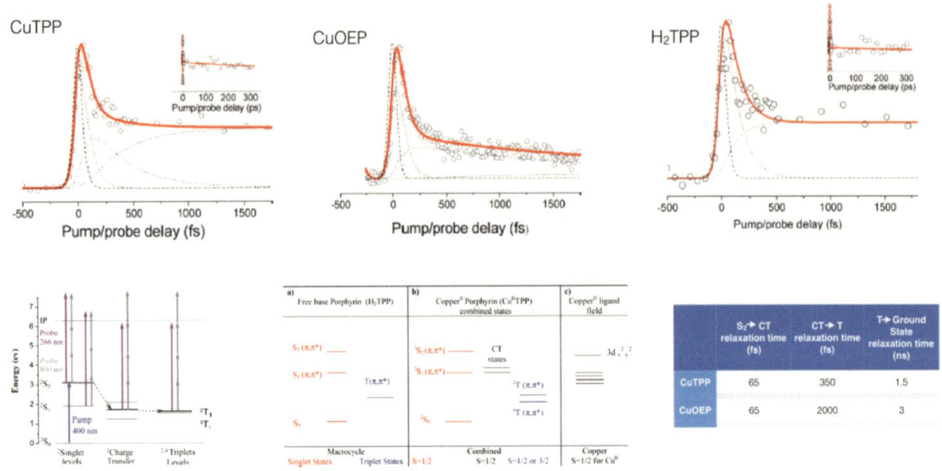

Figure 9. Top: CuTPP, CuOEP, and H_2TPP experimental pump-and-probe data. Bottom: (right) experiment energetics scheme, (middle) energy levels in CuTPP, CuOEP, and H_2TPP, and (left) table of excited states relaxation times. Adapted with permission from Ref. [49]. Copyright 2013, American Chemical Society.

remains similar to the porphyrin in solution demonstrating that the behavior of the molecular conformer in gas phase and in solution is similar.

As shown in the table of **Figure 9**, while the excitation dynamics in CuTPP and CuOEP presents the same decay time (65 fs) from the excited state S_2 to thecharge transfer (CT) state, the subsequent relaxations, first to the triplet states and finally to the ground state, are different: 350 fs and 2000 fs for CuTPP and CuOEP, and 1.5 ns and 3 ns, respectively. This is due to the different decay pathways in the two molecules, that is to the different back-bond configuration induced by the peripheral groups in the two cases (see Section 2.1).

The excitation dynamics data of the H_2TPP have to fit with the convolution of only two exponential curves because only two decay times, $S_2 \rightarrow S_1$ and $S_1 \rightarrow$ ground state, are present and their time constants are 110 fs and 2 ns, respectively. Clearly, in this case, the fast CT channel is missing because there is no metal atom in the middle.

Effects based on the different occupancies of the central metal 3*d* states have been measured in porphyrins [47], while in the Zn complexes, the decay time of the $S_2 \rightarrow$ CT shows long living excited states (600 fs); the analogous lifetimes in partially filled metal atoms result short of approximately one order of magnitude, up to about 50 fs in FeOEP.

It is quite evident that even the presence of one extra electron in the central atom can lead to different behavior in the relaxation dynamics processes.

Thick films (intermolecular interactions perturb the dynamics): In this case, metallo-phthalocyanine (CoPc) and metal-free Pc relaxation dynamics were investigated by means of pump-and-probe absorption spectroscopy in the UV-VIS spectral region [48]. The study revealed strong differences between the solution and thin films, with relevant dependence on the presence of the central metal atom. The samples have been resonantly optically excited (pump) in the Q band (>600 nm) and probed by a supercontinuum spectrum in the range of (1200-460) nm, that is the whole absorption spectrum of the system was probed as a function of the delay time.

The effects induced by aggregation are visible in the optical absorption spectra. The measured decay times of Co complexes result much faster with respect to the metal-free phthalocyanine. The coupling between the π orbitals and the 3*d* orbitals provides an additional channel of energy relaxation (see **Figure 10**), as found in porphyrins and previously discussed. Moreover, in films, the intermolecular interactions take place together with the excitonic processes. These differences are enlightened in the optical absorption data, where a clear broadening of the S and Q bands is induced by the molecular aggregation. Actually, in case of CoPc, the singly occupied $3d_z$, oriented orthogonal to the macrocycle plane, plays a major role. The related molecular orbital gives rise to delocalized exciton states that upon excitation result in resonance with the first excited state. This leads to an ultrafast charge transfer from the ligand excited state and the exciton. The absence of the 3*d* orbitals in the H_2Pc inhibits this process and only two decay times are detectable.

Recently, in a study on the correlation between dynamics and exciton length in phthalocyanine films [49], valuable results are reported on the dependence of relaxation dynamics on the film's thickness in planar ZnPc and TiOPc (the latter presents an additional ligand orthogonal to the macrocycle plane). By means of TR-PES, it has been proved that the exciton transport is consistent with ballistic or coherent transport mechanisms in ZnPc, while in TiOPc, the exciton transport mechanism is diffusive, due to the intramolecular hopping, absent in the first case.

As discussed previously, the molecular single layers deposited onto conductive substrates can lead to a plethora of bonds and molecule-substrate and intermolecular interactions that are specific of the systems. The dynamic study of the interface state's decay induced by the pump provides a detailed description of fundamental physical processes that govern the charge carrier dynamics at the interface.

As described in the previous section, TR-PES can provide information on the joint density of states evolution with time after the optical excitation, inducing for example HOMO-LUMO transition for selected pump energy. A nice example of TR-PES from molecular thin films of PCPDTBT deposited on ITO is presented in Ref. [48]. The measurements performed by

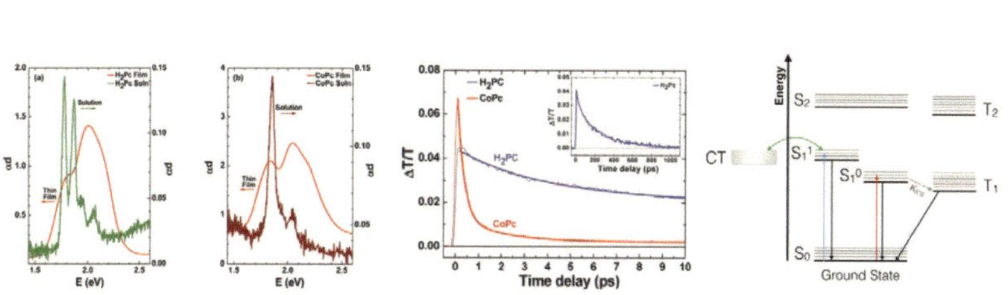

Figure 10. Left: Optical absorption spectra in film and solution for H_2Pc and CoPc, (middle) relaxation dynamics spectra for H_2Pc, and CoPc and (right) relaxation mechanisms for CoPc. Adapted with permission from Ref. [48]. Copyright 2010–2017, American Chemical Society.

pumping at the two absorption resonance bands, 400 nm (3eV) and 800 nm (1.5 eV), single out excitation of electrons from the HOMO to the LUMO orbitals as a function of the delay of the probe with respect to the pump, that is the filling and the subsequent de-excitation of the LUMO orbitals in the films.

Single Layers: The experimental study of single layer dynamics is quite challenging due to the very low signals involved. Only a few studies are present in the literature. The monolayer films onto coinage metal ordered surfaces can induce hybrid interface states in phthalocyanine systems. Harris and coworkers [50] compare the interfaces of H_2Pc and FePc on Ag(111). The unoccupied interface state residing just above the Fermi level is the fulcrum of the investigation. From the experimental data, there is evidence on the formation of an interface state (IS in **Figure 11**) formed through a hybridization process of π orbitals of the molecule and the surface state of the substrate. The formation of this hybrid interface state is slightly affected by the presence of the TM in the macrocycle that, as discussed in previous sections, chemically interacts with the substrate. Authors claim that the FePc/Ag(111) results suggest that the interaction between the metal center and the Ag(111) surface is not strong enough to fundamentally alter the hybridization phenomenon.

This example demonstrates how TR-PES provides useful information on the unoccupied interface state and hence on the hybridization process.

The role of the substrate orientation in the photo-induced electron dynamics at the porphyrin/Ag interface [51] has been proved by pump-and-probe TR-PES experiments.

Due to the energy level alignment of the ZnTPP porphyrin on Ag(100), the first excited state is formed through the Ag-unoccupied sp bands. This configuration favors an indirect charge transfer path, from the substrate to the molecule during the pump-and-probe process, opening an excitation channel 250 fs after the laser pump. The same time-resolved measurements carried out on porphyrin/Ag(111) show that in the latter case such an indirect path is not viable due to the different electronic configurations of the (111) interface.

In **Figure 12** right are evidenced the two different electronic configurations, while on the left, the excitation process is sketched.

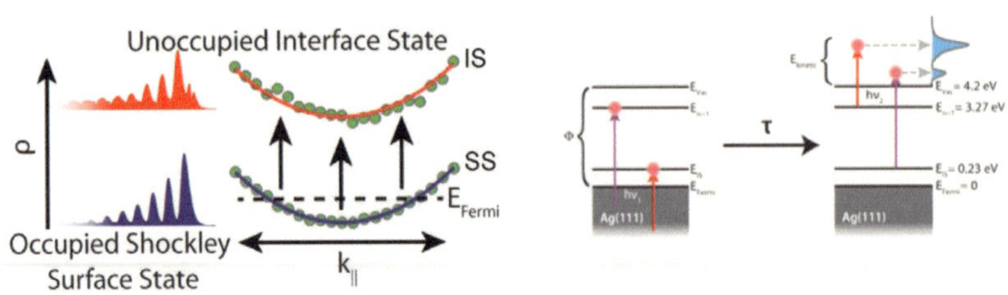

Figure 11. Left: Density of states for the occupied surface state and the unoccupied interface state and relative effective masses. Right: Schematic of the TR-PES process for H_2Pc/Ag(111): An electron is pumped into an intermediate state by an infrared photon, and then, after a time delay, it is photoemitted by a UV photon and its kinetic energy is measured. Adapted with permission from Ref. [50]. Copyright 2014, American Chemical Society.

http://dx.doi.org/10.5772/68111

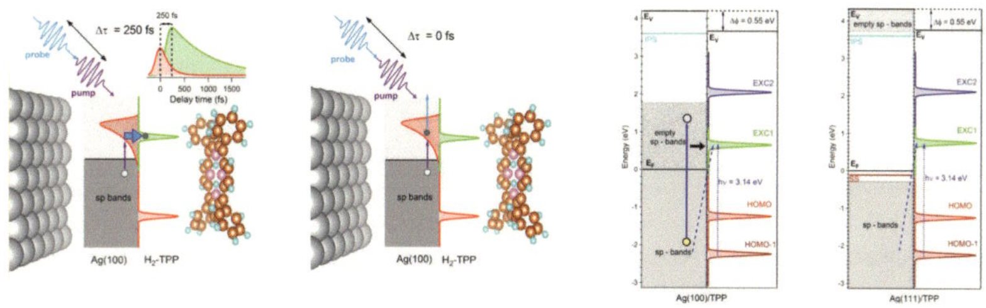

Figure 12. Left: An experimental set up of the pump-and-probe experiments. Right: A comparison of ZnTPP/Ag(100) and ZnTPP/Ag(111) electronic configurations. Adapted with permission from Ref. [51]. Copyright 2016, American Chemical Society.

The influence on the formation of the interface state, depending on the crystallographic orientation of the substrate, is also confirmed for other molecules, such as the PTCDA, by 2PPE experimental data [3], providing nice confirmation of the trend found by TR-PES.

Other important classes of experiments are related to TMC presenting spin cross-over processes in solution [52] and time-resolved X-ray absorption and diffraction [40]. Recently, time-resolved XPS core-level photoemission measurements performed on donor/acceptor blends of TMC organic molecules and fullerenes [53, 54] have shown that the core levels experience binding energy shifts after the systems are excited by a femtosecond optical pump.

5. Conclusions

Control of the electronic levels is crucial for optoelectronic applications. Optical and magneto-electronic properties are intimately connected, and the accurate determination of electronic levels and de-excitation dynamics is mandatory in the optimization of electronic, photovoltaic, and spintronic devices. We have shown the interconnection between structural, electronic properties, and the relaxation dynamics in a single layer of transition metal complexes.

The study of ordered layers and SAMs on single crystal substrates allows for a detailed investigation of fundamental physical processes that govern the charge carrier dynamics at the interface. Although these systems differ from more disordered aggregates that resemble metal-organic contacts of real devices, their dynamical characterization completed by a robust determination of the electronic and structural properties can shed light on the processes governing important applications.

Acknowledgements

The Italian Eurofel program and the CITIUS project financed through Interreg Italia-Slovenija, 2007–2013, are acknowledged. Professor Giovanni De Ninno is also acknowledged for support and fruitful discussion.

Author details

Maddalena Pedio[1]* and Barbara Ressel[2]

*Address all correspondence to: pedio@iom.cnr.it

1 CNR-IOM Laboratory TASC, Area Science Park, Basovizza, Trieste, Italy

2 University of Nova Gorica, Ajdovščina, Slovenia

References

[1] Geoffrey A. Lawrance. Introduction to coordination chemistry. John Wiley & Sons, United Kingdom; 2009–2010. ISBN 9780470519301, Online ISBN 9780470687123, doi:10.1002/9780470687123

[2] Chilukuri B, Mazur U, Hipps KW. Effect of dispersion on surface interactions of cobalt(II) octaethylporphyrin monolayer on Au(111) and HOPG(0001) substrates: a comparative first principles study. Phys. Chem. Chem. Phys. 2014;**16**:14096. doi:10.1 039/c4cp01762e

[3] Marks M, Schöll A, Höfer U. Formation of metal–organic interface states studied with 2PPE. J. Electr. Spectrosc. Relat. Phenom. 2014;**195**:263–271. doi:10.1016/j.elspec.2014.02.009

[4] Krause S, Casu MB, Schöll A, Umbach E. Determination of transport levels of organic semiconductors by UPS and IPS. New J. Phys. 2008;**10:**085001 (pp. 16). doi:10.1088/1367-2630/10/8/085001

[5] Koch N, Ueno N, Wee Andrew TS, editors. The molecule-metal interface. Weinheim: Wiley-VCH Verlag GmbH & Co KGaA; 2013. 255 p. doi:10.1002/9783527653171

[6] Zahn DRT, Gavrila GN, Salvan G. Electronic and vibrational spectroscopies applied to organic/inorganic interfaces. Chem. Rev. 2007;**107**:1161–1232. doi:10.1021/cr050141p

[7] DeLongchamp DM, Kline RJ, Fischer DA, Richter LJ, Toney MF. Molecular characterization of organic electronic films. Adv. Mater. 2010;XX:1–19.

[8] Gao Y. Surface analytical studies of interfaces in organic semiconductor devices. Mater. Sci. Eng. 2010;**68**:39–87. doi:10.1016/j.mser.2010.01.001

[9] Ishii H, Sugiyama K, Ito E, Seki K. Energy level alignment and interfacial electronic structures at organic/metal and organic/organic interfaces. Adv. Mater. 1999;**11**:605–625. doi:10.1002/(SICI)1521-4095(199906)11:8<605::AID-ADMA605>3.0.CO;2-Q

[10] Wang X-F, Tamiaki H. Cyclic tetrapyrrole based molecules for dye-sensitized solar cells. Energy Environ. Sci. 2010;**3**:94–106. doi:10.1039/b918464c

[11] Green J, De Cleva P. Photoionization cross-sections: a guide to electronic structure. Coord. Chem. Rev. 2005;**249**:209–228. doi:10.1016/j.ccr.2004.02.012

[12] Bidermane I, Brumboiu IE, Totani R, Grazioli C, Shariati-Nilsson MN, Herper HC, Eriksson O, Sanyal B, Ressel B, de Simone M, Lozzi L, Brena B, Puglia C. Atomic contributions to the valence band photoelectron spectra of metal-free, iron and manganese phthalocyanines. J. Electr. Spectrosc. Rel. Phenom. 2015;**205**:92–97. doi:10.1016/j.elspec.2015.09.004

[13] Gottfried JM. Surface chemistry of porphyrins and phthalocyanines. Surf. Sci. Rep. 2015;**70**:259–379.

[14] Auwärter W, Écija D, Klappenberger F, Barth JV. Porphyrins at interfaces. Nat. Chem. 2015;**7**:105–120. doi:10.1038/NCHEM.2159

[15] Sirringhaus H, et al. Two-dimensional charge transport in self-organized, high-mobility conjugated polymers. Nature. 1999;**401**:685–688. doi:10.1038/44359

[16] Sato Osamu. Dynamic molecular crystals with switchable physical properties. Nat. Chem. 2016;**8**:644–656. doi:10.1038/NCHEM.2547

[17] Senge MO, Ryan AA, Letchford KA, MacGowan SA, Mielke T. Chlorophylls, symmetry, chirality, and photosynthesis. Symmetry. 2014;**6**:781–843. doi:10.3390/sym6030781

[18] Gargiani P, Rossi G, Biagi R, Corradini V, Pedio M, Fortuna S, Calzolari A, Fabris S, Cezar JC, Brookes NB, Betti MG. Spin and orbital configuration of metal phthalocyanine chains assembled on the Au(110) surface. Phys. Rev. B. 2013;**87**:165407. doi:10.1103/PhysRevB.87.165407

[19] Bartolomé J, et al. Chapter 9: Magnetism of metal phthalocyanines. In J. Bartolomé, et al. editors, Molecular magnets. Heidelberg: Springer; 2014. ISBN 9783642406096

[20] Denys Nautschenko, M. Pedio unpublished

[21] Krasnikov SA, Preobrajenski AB, Sergeeva NN, Brzhezinskaya MM, Nesterov MA, Cafolla AA, Senge MO, Vinogradov AS. Electronic structure of Ni(II) porphyrins and phthalocyanine studied by soft X-ray absorption spectroscopy. Chem. Phys. 2007;**332**:318–324. doi:10.1016/j.chemphys.2006.12.015

[22] Miedema PS, de Groot FMF. The iron L edges: Fe 2p X-ray absorption and electron energy loss spectroscopy. J. Electr. Spectr. Rel. Phenom. 2013;**187**:32–48. doi:10.1016/j.elspec.2013.03.005

[23] Jurow M, Schuckman AE, Batteas JD, Drain CM. Porphyrins as molecular electronic components of functional devices. Coord. Chem. Rev. 2010;**254**:2297–2310. doi:10.1016/j.ccr.2010.05.014

[24] Rossi G, d'Acapito F, Amidani L, Boscherini F, Pedio M. Local environment of metal ions in phthalocyanines: K-edge X-ray absorption spectra. Phys. Chem. Chem. Phys. 2016;**18**:23686–23694. doi:10.1039/c6cp04022e

[25] Baerends EJ, Ricciardi G, Rosa A, van Gisbergen SJA. A DFT/TDDFT interpretation of the ground and excited states of porphyrin and porphyrazine complexes. Coord. Chem. Rev. 2002;**230**:5–27. doi:10.1016/S0010-8545(02)00093-0

[26] Kroll T, Kraus R, Schönfelder R, Aristov V Yu, Molodtsova O V, Hoffmann P, Knupfer M. Transition metal phthalocyanines: Insight into the electronic structure from soft x-ray spectroscopy J. Chem. Phys. 2012;**137**:054306. doi: 10.1063/1.4738754

[27] Umari P, Fabris S. Importance of semicore states in GW calculations for simulating accurately the photoemission spectra of metal phthalocyanine molecules. J. Chem. Phys. 2012;**136**:174310 (6). doi:10.1063/1.4705360

[28] Nobuo Ueno "Electronic Structure of Molecular Solids: Bridge to the Electrical Conduction" 2012; pp. 65–89 in Physics of Organic Semiconductors, Second Edition. Wiley-VCH Verlag GmbH & Co KGaA; doi: 10.1002/9783527654949.ch3

[29] Djuric T, Ules TH, Gusenleitner S, Kayunkid N, Plank H, Hlawacek G, Teichert C, Brinkmann M, Ramsey M, Resel R. Substrate selected polymorphism of epitaxially aligned tetraphenyl-porphyrin thin films. Phys. Chem. Chem. Phys. 2012;**14**:262–272. doi:10.1039/c1cp22299f

[30] Marsili M, Umari P, Di Santo G, Caputo M, Panighel M, Goldoni A, Kumar M, Pedio M., Solid state effects on the electronic structure of H_2OEP, Phys. Chem. Chem. Phys. 2014;**16**:27104–27111.

[31] Stoll LK, Zgierski MZ, Kozlowski PM. Density functional theory analysis of nickel octaethylporphyrin ruffling. J. Phys. Chem. A. 2002;**106**:170–175.

[32] Ryuzaki S, Hasegawa T, Onoe J. , Effects of inter-molecular charge-transfer excitons on the external quantum efficiency of zinc-porphyrin/C60 heterojunction photovoltaic cells, J. Appl. Phys. 2009;**105**:113529.

[33] Schreiber F. Structure and growth of self-assembling monolayers. Progr. Surf. Sci. 2000;**65**:151–256.

[34] Resta A, Felici R, Kumar M, Pedio M. Ni and Cu octaethyl porphyrins ordered monolayer on Au(111) surfaces. J. Non-Cryst. Solids. 2010;**356**:1951–1954. doi:10.1016/j.jnoncrysol.2010.05.051

[35] Hipps KW. Scanning Tunneling Spectroscopy. In *Handbook of applied solid state spectroscopy*, D.R. Vij, editor. New York: Springer-Verlag; 2006. ISBN 0-387-32497-6

[36] Pedio M, Cepek C, Felici R. Organic molecules on noble metal surfaces: the role of the interface. Dr. Yen-Hsun Su, editor. Noble metals. InTech; 2012. doi: 10.5772/34551. Retrieved from http://www.intechopen.com/books/noble-metals/organic-molecules-on-noble-metal-surfaces-the-role-of-the-interface

[37] Koch N, Gerlach A, Duhm S, Glowatzki H, Heimel G, Vollmer A, Sakamoto Y, Suzuki T, Zegenhagen J, Rabe JP, Schreiber F. Adsorption-induced intramolecular dipole: correlating molecular conformation and interface electronic structure. J. Am. Chem. Soc. 2008;**130**:7300–7304. doi:10.1021/ja800286k

[38] Lodi Rizzini A, Krull C, Mugarza A, Balashov T, Nistor C, Raoul Piquerel R, Klyatskaya S, Ruben M, Sheverdyaeva PM, Moras P, Carbone C, Stamm CH, Miedem PS, Thakur PK, Sessi V, Soares M, Yakhou-Harris F, Cezar JC, Stepanow S, Gambardella P. Coupling of single, double, and triple-decker metal-phthalocyanine complexes to ferromagnetic and antiferromagnetic substrates. Surf. Sci. 2014;**630**:361–374. doi:10.1016/j.susc.2014.07.008

[39] Gargiani P, Angelucci M, Mariani C, Betti MG. Metal-phthalocyanine chains on the Au(110) surface: interaction states versus *d*-metal states occupancy. Phys. Rev. B. 2010;**81**:085412. doi: 10.1103/PhysRevB.81.085412

[40] see for example Cammarata M, Bertoni R, Lorenc M, Cailleau H, Di Matteo S, Mauriac C, Matar SF, Lemke H, Chollet M, Ravy S, Laulhé C, Létard J-F, Collet E. Sequential activation of molecular breathing and bending during spin-crossover photoswitching revealed by femtosecond optical and x-ray absorption spectroscopy. Phys. Rev. Lett. 2014;**113**:227402. doi: and refs therein

[41] https://alumni.stanford.edu/get/page/magazine/article/?article_id=39117

[42] Hentsche M, Kienberger R, Spielmann CH, Reider GA, Milosevic N, Brabec T, Corkum P, Heinzmann U, Drescher M, Krausz F. Attosecond metrology. Nature. 2001;**414**:509–513. doi:10.1038/35107000

[43] Pfeifer T, Spielmann C, Gerber G. Femtosecond x-ray science. Rep. Prog. Phys. 2006; **69**:443–505. doi:10.1088/0034-4885/69/2/R04

[44] Grazioli C, Callegari C, Ciavardini A, Coreno M, Frassetto F, Gauthier D, Golob D, Ivanov R, Kivimäki A, Mahieu B, Bucar B, Merhar M, Miotti P, Poletto L, Polo E, Ressel B, Spezzani C, De Ninno G. CITIUS: an infrared-extreme ultraviolet light source for fundamental and applied ultrafast science. Rev. Sci. Instr. 2014;**85**:023104. doi:10.1063/1.4864298

[45] Cappel UB, Plogmaker S, Terschlüsen JA, Leitner T, Johansson EMJ, Edvinsson T, Sandell A, Karis O, Siegbahn H, Svensson S, Mårtensson N, Rensmo H, Söderström J. Electronic structure dynamics in a low bandgap polymer studied by time-resolved photoelectron spectroscopy. Phys. Chem. Chem. Phys. 2016:**18**;21921–21929. 2016. doi:10.1039/C6CP04136A

[46] Ha-Thi MH, Shafizadeh N, Poisson L, Soep B. An efficient indirect mechanism for the ultrafast intersystem crossing in copper porphyrins. J. Phys. Chem. A., 2013;**117**: 8111–8118. doi:10.1021/jp4008015

[47] Shafizadeh N, Ha-Thi M-H, Poisson L, Soep B. Ultrafast electronic relaxation of excited state of biomimetic metalloporphyrins in the gas phase. In Dr. Lilyana Pramatarova, editor. On biomimetics. InTech; 2011. doi:10.5772/19976. Retrieved from http://www.intechopen.com/books/on-biomimetics/ultrafast-electronic-relaxation-of-excited-state-of-biomimetic-metalloporphyrins-in-the-gas-phase

[48] Gadalla A, Crégut O, Gallart M, Hönerlage B, Beaufrand J-B, Bowen M, Boukari S, Beaurepaire E, Gilliot P. Ultrafast optical dynamics of metal-free and cobalt phthalocyanine thin films. J. Phys. Chem. C. 2010;**114**: 4086–4092. doi:10.1021/jp911438y

[49] Wang T, Kafle TR, Kattel B, Chan W-L. Observation of an ultrafast exciton hopping channel in organic semiconducting crystals. J. Phys. Chem. C. 2016;**120**:7491–7499. doi:10.1021/acs.jpcc.6b01400

[50] Caplins BW, Suich DE, Shearer AJ, Harris JB. Metal/phthalocyanine hybrid interface states on Ag(111). J. Phys. Chem. Lett. 2014;**5**:1679–1684. doi:10.1021/jz500571z

[51] Tognolini S, Ponzoni S, Sedona F, Sambi M, Pagliara S. Role of the substrate orientation in the photoinduced electron dynamics at the porphyrin/Ag interface. J. Phys. Chem. Lett. 2015;**6**:3632–3638. doi:10.1021/acs.jpclett.5b01528

[52] Chergui M. On the interplay between charge, spin and structural dynamics in transition metal complexes. Dalton Trans. 2012;**41**:13022. doi:10.1039/c2dt30764b

[53] Arion T, Neppl S, Roth F, Shavorskiy A, Bluhm H, Hissain Z, Gessner O, Eberhardt W., Site-specific probing of charge transfer dynamics in organic photovoltaics Appl. Phys. Lett. 2015;**106**:121602. doi: 10.1063/1.4916278

[54] Ozawa K, Yamamoto S, Yukawa R, Akikubo K, Emori M, Sakama H and Matsuda I, Capturing transiently charged states at the $C_{60}/TiO_2(110)$ interface by time-resolved soft X-ray photoelectron spectroscopy Org. Electron., 2016;**31**:98–103. doi: 10.1016/j.orgel.2016.01.020